*Know-It-All*

# Energy

# *Know-It-All* Energy

**The 50 Most Elemental Concepts in Energy, Each Explained in Under a Minute >**

Editor **Brian Clegg**
Foreword **Jim Al-Khalili**

Contributors

**Philip Ball**
**Brian Clegg**
**Leon Clifford**
**Simon Flynn**
**Sharon Ann Holgate**
**Andrew May**

WELLFLEET
PRESS

First published in 2017 by Wellfleet Press,
an imprint of The Quarto Group,
142 West 36th Street, 4th Floor,
New York, New York 10018, USA
**T** (212) 779-4972 **F** (212) 779-6058
**www.QuartoKnows.com**



10 9 8 7 6 5 4 3 2 1

ISBN: 978-1-57715-161-6

This book was conceived, designed, and produced by
**Ivy Press**
An imprint of The Quarto Group
The Old Brewery, 6 Blundell Street
London N7 9BH, United Kingdom
**T** (0)20 7700 6700 **F** (0)20 7700 8066

Publisher **Susan Kelly**
Creative Director **Michael Whitehead**
Editorial Director **Tom Kitch**
Art Director **James Lawrence**
Project Editors **Jamie Pumfrey and Jenny Campbell**
Designer **Ginny Zeal**
Illustrator **Steve Rawlings**

Printed in China

# CONTENTS

# FOREWORD

## Jim Al-Khalili

There is no other term in the whole of science that has been so abused, misunderstood, or misused as the term "energy." We all think we understand what it means—after all, it pervades so much of our lives and everyday language that it has long since lost any sense of mystery, unlike many other common scientific concepts, such as "time," "space," or "mass," which still retain a certain inscrutability and dignity.

Another issue is that the notion of energy is so hard to pin down—it seems to be a catchcall term that covers what, at first glance, appear to be quite unrelated concepts. This is because energy does indeed come in many different forms. On the one hand, we can think of it as something physical, something we can see (light) and feel (heat). But energy is also the pull of the earth's gravity, the stored promise of a compressed spring, and the *élan vital* that differentiates the living from the inanimate. Everything in the world that moves, or changes, or interacts does so because it is making use of some form of energy.

A frustration to many scientists is that some people unwisely use the term "energy" as a vague metaphysical presence, or simply as a psychological feeling. They may say something such as "I could feel the negative energy as soon as I entered the room" or "he exuded positive energy." Such misguided notions serve only to muddy the water farther.

But energy can be defined. We just need to understand the rules of the game. For example, when a scientist or engineer uses the term "work" as a scientific concept, they mean something quite precise—it is an action on a system that transfers energy from one form to another, or from one place to another. Indeed, the entire discipline of thermodynamics explores the relationship between heat, work, and energy. And this doesn't even touch on nuclear energy, or the chemical energy in our cells, or the nature of dark energy out in space.

Like others in the *Know-It-All* series, this book will give you a delightfully fresh perspective on energy, in all its forms and in bite-sized chunks, by some of the best explainers around.

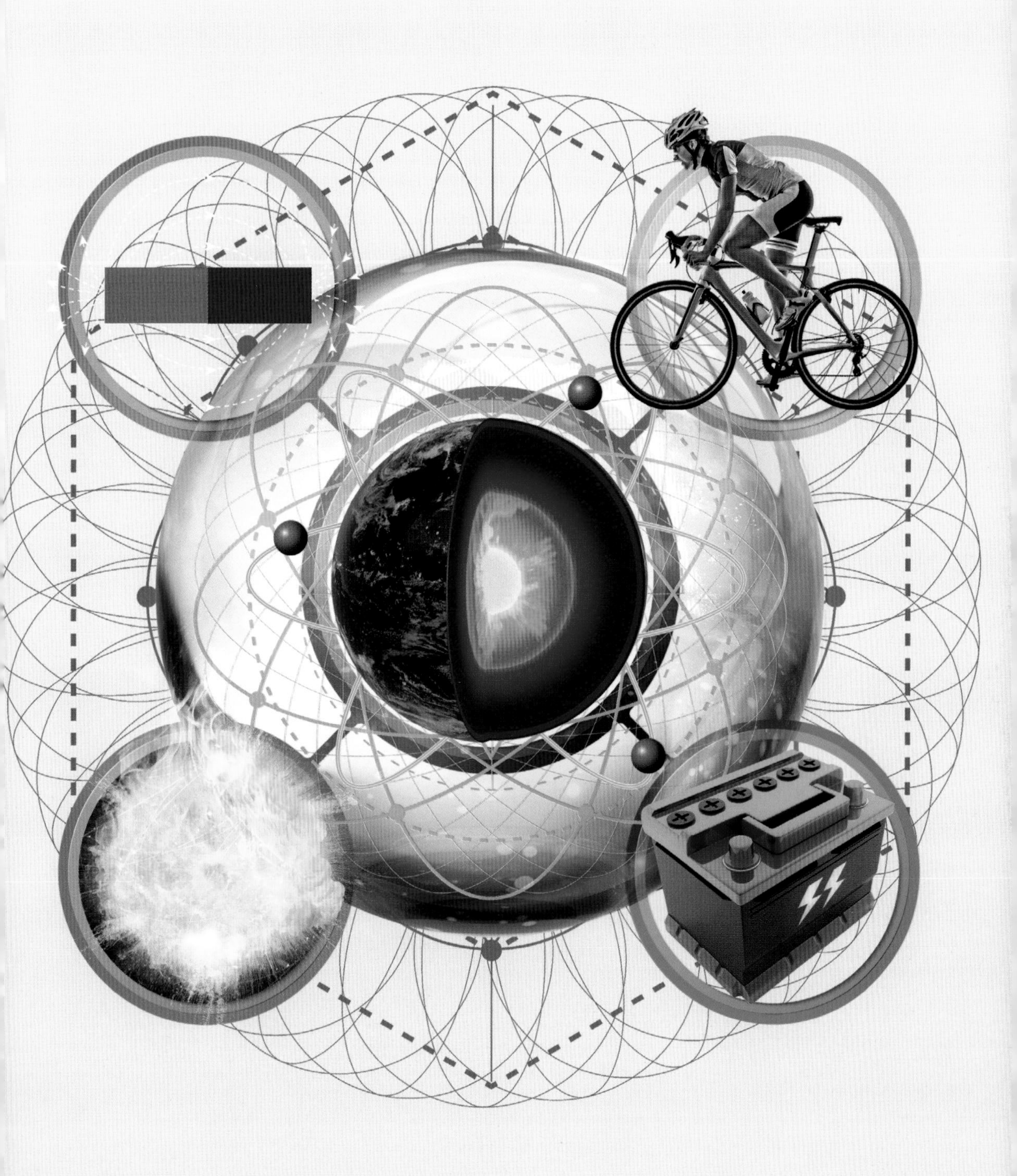

# INTRODUCTION
Brian Clegg

It is practically impossible to pin down exactly what energy is, but far easier to say what it does—energy makes things happen. Whether we're thinking of machinery, living things, or the flow of a stream down a mountainside, there's energy at work. And as Einstein's equation $E=mc^2$ shows, we now know that energy is not just what makes things happen to matter—energy and matter are interchangeable. Energy is the vital stuff of the universe.

Because we were slow to realize how many guises energy takes, it has traditionally been divided into many forms—although the distinctions between these different forms are not as clear as they once were. For example, when atoms or molecules are moving in a substance we can't separate their heat energy from the kinetic energy of their movement—it can still be useful to think of those different "kinds" of energy. For this

reason we will begin our exploration by classifying energy from heat, through potential and chemical energy to nuclear and mass energy. All this is pulled together in the overarching discovery that unites these different ways that energy shows itself. While we can transform energy, we can neither make it nor destroy it—energy is conserved.

The modern world is so energy-rich and energy-dependent that a mention of energy outside the scientific context will usually make us think of electric power generation. But with the basics under our belts, it makes sense first to consider energy in nature. We don't make our own energy, we transform it from natural sources, whether it's the gravity that lies behind a hydroelectric plant, the chemical energy stored by living things that we use in fossil fuel, or the vast natural nuclear reactor that is the Sun (which gives us, directly or indirectly, the clear majority of the energy we use).

## Storing and Transmitting Energy

If we think of all that energy being passed around—from sunlight to plants, and from plants to other living things in our power stations and vehicles—it becomes obvious that to make such an energy economy work we need the equivalent of banks and wallets. Energy storage is a fundamental requirement to make sensible use of energy. We find this in nature, where the energy from sunlight and eating is stored away in the structure of tiny molecules called ATP. Our complex networks of electricity also drive us to store electrical energy. The difficulty of developing better battery technology, for example, is probably the single biggest barrier to a truly clean energy future.

In most cases energy isn't stored up where it's going to be used, and so it needs to be transmitted from place to place and converted from one form to another. Some of the most exciting technical developments in recent years, from the use of lasers to superconductors in various projects, have been involved in energy transmission. At the same time, we have become aware of the impact our energy habits are having on the planet and are dedicating increasingly large efforts to green mechanisms in order to minimize that effect.

After exploring the wide-ranging impact of energy, we bring the book to a close with an associated concept—entropy—that is so tightly intertwined with energy that it is impossible to separate the two. Entropy, a measure of order and disorder, is at the heart of why we can't make a perpetual motion machine, or an engine that perfectly uses all the energy available. Entropy could even forecast the end times of the universe and so gives us the ideal close on our picture of this most universal of concepts.

## How This Book Works

Each topic is clearly and concisely explained on one page in a punchy single paragraph. For an even quicker overview, there is the **3-second thrash**—the key idea caught in a single sentence. Then the **3-minute thought** expands on this, addressing the consequences of a theory or drawing out a quirky, intriguing aspect of the subject. Each chapter also contains the profile of a pioneer in energy theory or applications—people such as James Prescott Joule, Alessandro Volta, and William Thomson, Lord Kelvin—and a glossary that explains key terms and trickier concepts.

# THE BASICS

## THE BASICS
GLOSSARY

**antimatter** Matter in which some properties, such as electrical charge, are reversed. Colliding equivalent matter and antimatter particles annihalate to produce pure energy.

**atoms** Smallest matter particles constituting a chemical element. All atoms jiggle about, while those in liquids and gases also travel through the fluid. The combined effect of this motion is heat.

**conservation of energy** Several phenomena in nature are conserved, including energy. This means that in a system isolated from the outside, the energy stays the same. Because $E=mc^2$ shows that mass and energy are interchangeable, we need to include mass and energy.

**$E=mc^2$** Einstein's equation describes the relationship between energy, $E$, and the mass of a substance, $m$, linked by the square of the speed of light, $c$.

**electromagnetic radiation** Moving electricity produces magnetism, while moving magnets produce electricity. At the right speed a moving wave of electricity will generate magnetism, which generates electricity and so on: This is electromagnetic radiation, which includes radio, microwaves, light, and X-rays.

**electrostatic bonds** When atoms join to make molecules, the electrical attraction between negatively charged electrons on one atom and the positive charge of the other atom is called electrostatic bonding.

**electrons** Negatively charged fundamental particles making up the outer part of an atom. Electrons act as carriers of electrical energy.

**entropy** Term describing the level of disorder in an object or collection of objects, measured as the number of different ways the components of the object can be arranged.

**enzymes** Complex molecules produced in living organisms; enzymes are catalysts—substances that speed up a chemical reaction but don't form part of the reaction's final products.

**free energy** Energy in a system available to do work. In a chemical reaction, the free energy is the difference between the energy of the initial and final states of the atoms or molecules taking part.

**fundamental conservation laws** In physics, several quantities stay the same in an isolated system. These include mass, or energy, momentum, and angular momentum (a measure of the spin of a body).

**gravitational field** A field describes the value of anything at every point in space (and time). In physics, fields are used to describe, for example, the strength of gravitational attraction.

**gravity well** The gravitational attraction of a body—for example, the Earth—gets bigger as we get closer to it. To escape from that body's attraction takes energy, which is likened to getting out of a well, hence the term "gravity well."

**molecules** Atoms link because of the attraction between negative and positive electrical charges. Such combined atoms are called molecules, which can vary from a simple molecule combining two identical atoms to the complex molecules of DNA.

**momentum** Mass of an object multiplied by its velocity. Momentum is a measure of the amount of "oomph" a moving body has.

**protons and neutrons** Heaviest particles in an atom. Protons (which have a positive charge) and neutrons (which have no electrical charge) form a tight core called the nucleus.

**strong nuclear force** Protons and neutrons are made of smaller particles called quarks. The strong nuclear force keeps quarks together to make up the bigger particles. It also holds the protons and neutrons together in the nucleus despite the repulsion of positively charged protons.

**theory of relativity** Einstein's extension of Newton's laws of motion to include light's constant velocity. The result is that it is not possible to separate space and time, as movement influences the passage of time.

**thermodynamics** Literally "the movement of heat." Thermodynamics is the science of how heat energy is transferred from place to place, central to understanding any process involving heat.

**velocity** Combines speed and direction of movement, for example 98 ft/sec (30 m/sec) horizontally.

# WHAT IS ENERGY?

**3-SECOND THRASH**
Energy comes in many interconvertible forms, and can be harnessed to do useful work. But it's hard to say what it *is*.

**3-MINUTE THOUGHT**
We don't know how much energy our universe contains, but it might be none. It's perfectly possible, in theory, to create a universe with zero energy, and ours might be that way because all the positive energy in light, heat, and matter could be perfectly balanced by the "negative energy" bound up in gravitational fields. If this were the case, the universe should be uniform on average—as, indeed, it seems to be.

This is a somewhat unusual confession in a book of this nature, so we shall let the great physicist Richard Feynman make it: "It is important to realize that in physics today, we have no knowledge of what energy *is*." As Feynman explains, we know that energy comes in many interconvertible forms, and we can say something about each of them. But as to what energy as a raw concept is, we struggle to give an answer. We might say that energy is something we can use to do useful work, like lifting up an object—except that some energy can't be used in that way. Chemical energy can be stored in fuels and released in chemical reactions. Kinetic energy is possessed by moving bodies. Gravitational potential energy is something an object has by virtue of its position in a gravitational field, for instance, high above the ground. Warm objects have heat energy. But what it is that all these forms of energy share, apart from the fact that they can be converted from one form to another? It's hard to say. It is almost like a vital essence that drives change and movement in the universe—but there's nothing mystical in that. Energy, like life or art, is best understood by example, not by definition.

**RELATED TOPICS**
See also
HEAT
page 18

CONSERVATION OF ENERGY
page 32

**3-SECOND BIOGRAPHIES**
GOTTFRIED WILHELM LEIBNIZ
1646–1716
German philosopher who introduced a scientific notion akin to (kinetic) energy, called *vis viva*

THOMAS YOUNG
1773–1829
English polymathic scientist who was the first person known to use the word "energy" in a scientific context in 1802

**EXPERT**
Philip Ball

***Energy is involved in practically everything we experience—from the output of the Sun to our ability to move—and can be transformed to perform a variety of different functions.***

# HEAT

We know what it implies for something to be hot, but the form of energy responsible for it—heat—is a trickier concept than it might appear. Heat is a form of energy associated with the motion of an object's atoms and molecules: The more vigorously they jiggle about, the more heat the object contains. But heat is not exactly identical to that motional (kinetic) energy of particles. It refers to a transfer of energy, from a hot body to a cooler one. It's something that flows, and indeed was once considered a sort of fluid. If some of the internal energy due to atomic motions in one body is transferred to the internal energy of another, heat has flowed between them. This can happen via direct physical contact—heat conduction—or via the emission of electromagnetic radiation, such as light or infrared, through space: heat radiation. Heat, or thermal energy, is often considered "low-grade energy" because it can be hard to harness to do work. During energy transfers, some of it almost inevitably ends up being wasted as heat, which is dispersed into the environment—for example, through friction, which causes heating in electrical wires and devices and which dissipates the kinetic energy of flowing fluids.

**3-SECOND THRASH**
Heat is a flow of energy caused by differences in internal energy: it flows from hot to cold, making the atoms of the cooler body jiggle more vigorously.

**3-MINUTE THOUGHT**
Because some energy is always "wasted" when harnessing it for work, generally ending up as heat dispersed inaccessibly throughout the surroundings, it seems inevitable that eventually all energy in the universe will end up in thermal form, and the heat will spread out evenly. Then it will be impossible to recover any energy to perform work, and nothing can happen. This situation, first discussed in the mid-nineteenth century, became known as the heat death of the universe.

**RELATED TOPICS**
See also
KINETIC ENERGY
page 20

CHEMICAL ENERGY
page 26

THE SECOND LAW
page 142

**3-SECOND BIOGRAPHIES**
ANTOINE LAVOISIER
1743–94
French chemist who imagined heat as a "subtle" but tangible fluid called caloric, and devised instruments (calorimeters) to measure it

BENJAMIN THOMPSON, COUNT RUMFORD
1753–1814
American-British scientist whose experiments on artillery led him to develop a theory of heat caused by motion and friction

**EXPERT**
Philip Ball

***Heat energy is present in many processes, whether unwanted or to produce an effect such as melting metal.***

# KINETIC ENERGY

**3-SECOND THRASH**
Kinetic energy is associated with the motion of an object; its value rises in proportion to the mass of the object and the square of its velocity.

**3-MINUTE THOUGHT**
Einstein's famous equation $E=mc^2$ is usually interpreted in terms of the "rest mass" energy associated with a stationary object. However, the equation is also true in the case of a moving object, for which the kinetic energy must be added to the rest mass energy. Since the speed of light, $c$, is constant, this means the effective mass, $m$, of the object must increase as its speed increases.

"Kinetic" comes from the Greek word for motion, and the kinetic energy of an object is simply the energy it possesses by virtue of its motion. The amount of kinetic energy in joules is equal to half the object's mass in kilograms, multiplied by the square of its speed in meters per second. Thus a small, fast-moving object can have as much kinetic energy as a massive, slow-moving one. A 50-ton tank traveling at 32.8 ft/s (10 m/s) has a kinetic energy of 2.5 million joules, but so does an 11 pound (5 kg) shell fired at 3,280 ft/sec (1,000 m/s) from the tank's gun. Explosive energy is another form of kinetic energy, deriving from the motion of the blast fragments flying off in all directions. All types of motion contribute to an object's kinetic energy, not just motion in a straight line. For example, there is kinetic energy associated with the rotation of an electric motor, with the back-and-forth movement of the pistons in a car engine and with the random motions of atoms and molecules. In this last case, kinetic energy on a microscopic scale is perceived as heat on a macroscopic scale. In fact, the temperature of an object is simply a measure of the average kinetic energy of its constituent molecules.

**RELATED TOPICS**
See also
HEAT
page 18

POTENTIAL ENERGY
page 22

CONSERVATION OF ENERGY
page 32

**3-SECOND BIOGRAPHIES**
GASPARD DE CORIOLIS
1792–1843
French physicist who derived the $\frac{1}{2}mv^2$ formula for the energy of motion

WILLIAM THOMSON, LORD KELVIN
1824–1907
Scottish physicist who coined the term "kinetic energy" and showed its relation to temperature

**EXPERT**
Andrew May

***Although a tank shell has far less mass than a tank, it has the same kinetic energy as it travels at a much higher velocity.***

# POTENTIAL ENERGY

**3-SECOND THRASH**
Potential energy is energy that is inherent in an object, for example due to its position in a gravitational field or the structure of its chemical bonds.

**3-MINUTE THOUGHT**
Potential energy (PE) is only defined in relative terms, so the point at which it is zero can be chosen wherever it is most convenient. For example, aeronautical engineers usually think of PE as being zero at ground level and increasing with increasing altitude. Astronomers, on the other hand, tend to think of PE as a negative quantity, which eventually rises to zero at an infinite distance from the Earth. This may sound bizarre, but it simplifies many calculations.

An apple falling from a tree possesses kinetic energy (KE) from its movement. Yet nothing gave it that energy. It wasn't pushed off the tree, so it must already have had the same energy when it was hanging on the branch. This is an example of potential energy (PE)—in this case gravitational potential energy, due to the apple's elevated position in the Earth's gravitational field. Near to the Earth's surface, gravitational potential energy is proportional to an object's mass and its height above ground; at greater distances the formula is more complicated. One of the reasons spacecraft require such high launch velocities is that they need to trade kinetic energy for potential energy as they climb up out of Earth's gravity well. Electric fields also give rise to potential energy, in an analogous way to gravitational fields. This is particularly important on molecular scales, because it means there is potential energy locked up in the electrostatic bonds between atoms. The energy released during a chemical reaction—for example, when fuel is burned or food is metabolized—comes from this binding energy, which is higher in the initial configuration of its molecules than the final one. On a more powerful scale, the same is true of nuclear reactions. Thus both chemical energy and nuclear energy are ultimately different forms of potential energy.

**RELATED TOPICS**
See also
CHEMICAL ENERGY
page 26

CONSERVATION OF ENERGY
page 32

GRAVITY
page 38

**3-SECOND BIOGRAPHIES**
DANIEL BERNOULLI
1700–82
Swiss mathematician who recognized that the sum of KE and PE remains constant

WILLIAM RANKINE
1820–72
Scottish engineer who coined the term "potential energy" in the context of steam engines

**EXPERT**
Andrew May

***As a coin drops from a high building, its potential energy due to its position in the Earth's gravitational field is converted into kinetic energy of movement.***

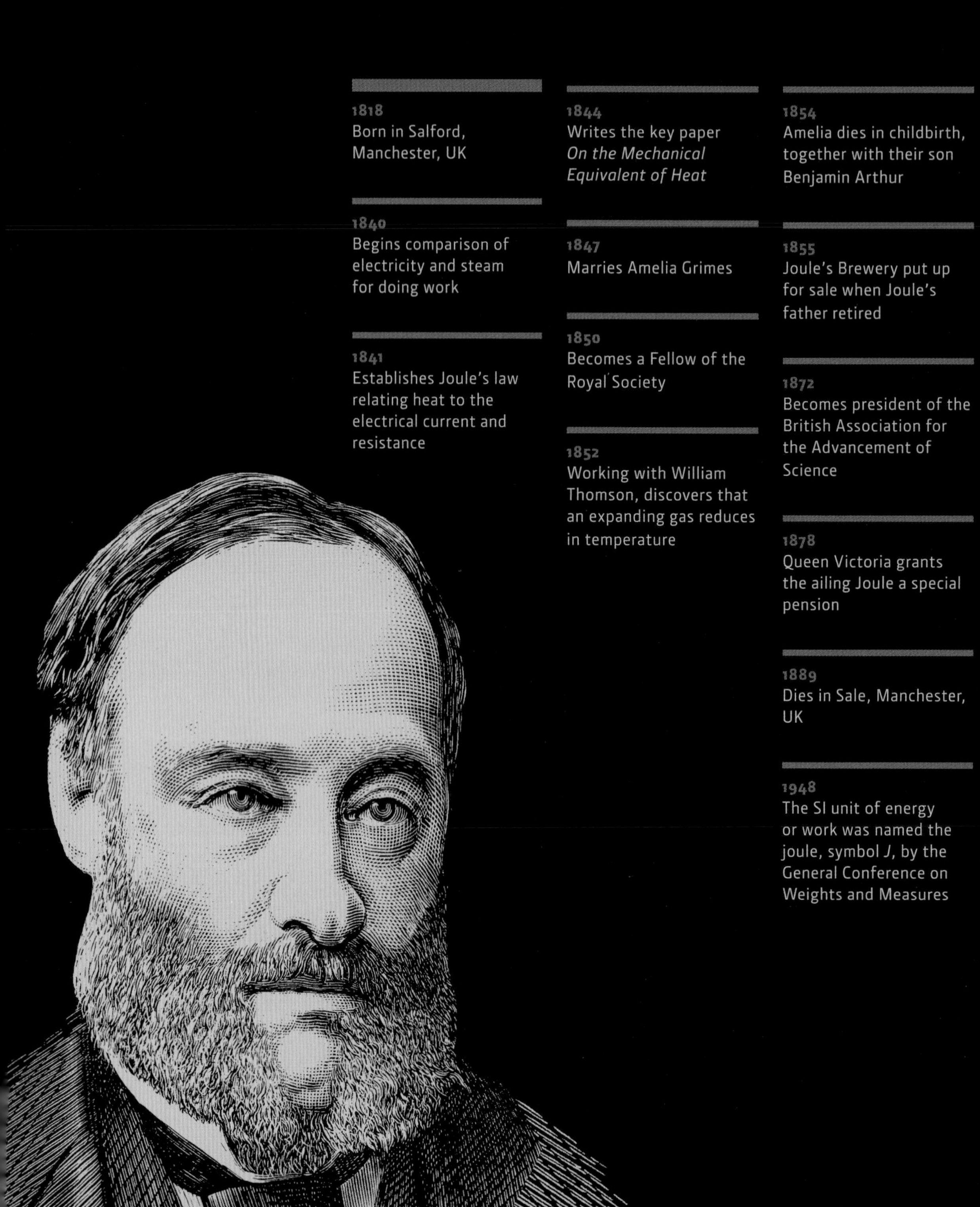

1818
Born in Salford, Manchester, UK

1840
Begins comparison of electricity and steam for doing work

1841
Establishes Joule's law relating heat to the electrical current and resistance

1844
Writes the key paper *On the Mechanical Equivalent of Heat*

1847
Marries Amelia Grimes

1850
Becomes a Fellow of the Royal Society

1852
Working with William Thomson, discovers that an expanding gas reduces in temperature

1854
Amelia dies in childbirth, together with their son Benjamin Arthur

1855
Joule's Brewery put up for sale when Joule's father retired

1872
Becomes president of the British Association for the Advancement of Science

1878
Queen Victoria grants the ailing Joule a special pension

1889
Dies in Sale, Manchester, UK

1948
The SI unit of energy or work was named the joule, symbol *J*, by the General Conference on Weights and Measures

# JAMES PRESCOTT JOULE

"Joule" is probably more familiar as the unit of energy than a man. In many ways, James Joule came from the tradition of engineers like James Watt, pragmatically looking for ways to improve their technology. Joule's father, Benjamin, was a successful brewer who paid for his son to be tutored by the Manchester-based originator of modern atomic theory, John Dalton. James Joule remained a brewer first, joining the management of the brewery in his early 20s, and a scientist second. Like any industrialist who employed engines, he wanted to get the most for his money and began a study of the amount of work different sources of energy could produce.

Work in the physics sense is just energy being moved around. Joule was interested to discover if the new electric motors could provide a more efficient solution than the steam engines employed in the Joule's Brewery. In experimenting with electricity, he came up with his first major contribution, Joule's law, that the heat produced by a current in a wire is proportional to the square of the current multiplied by the resistance of the wire. From his viewpoint, the key outcome of his electrical work was the discovery that coal was more economical than electricity.

Putting electricity and heat alongside each other inspired him to think about how different forms of energy are connected. At the time, heat was considered to be the flow of a special fluid called caloric, which moved from hot to cold objects, and could not be made or destroyed. But Joule was increasingly of the opinion that the energy of electricity or mechanical work could be converted into heat. He put together a range of experiments to demonstrate this, most elegantly with a device where a falling weight, producing mechanical work from potential energy, turned a paddle wheel in a container of water, increasing the temperature of the water.

Building on his training with Dalton, Joule reasoned that such heating could only happen if the heat of a body reflected the kinetic energy of the molecules in it. At the time this was controversial, with the existence of atoms and molecules still in doubt, especially as Joule's theory meant that the particles had to bounce off each other without losing any energy, something that never happened with ordinary objects. Many rejected Joule's ideas, but now the Salford brewer is considered a key figure in understanding energy.

***Brian Clegg***

# CHEMICAL ENERGY

**3-SECOND THRASH**
Chemical energy is stored in molecules and released by changes in the ways their atoms are linked and arranged.

**3-MINUTE THOUGHT**
Some of the most energy-dense chemical compounds known are high explosives such as TNT, RDX, and PETN—the latter two are ingredients of Semtex. They are substances that release their chemical energy in a fast, supersonic outburst. Typically they contain oxygen atoms to promote rapid combustion of carbon-rich parts, as well as nitrogen atoms, which release a lot of energy when they combine in pairs as nitrogen molecules, with very strong bonds.

We are chemically powered, and so, on the whole, is the world we have made. Our bodies run on chemical energy harvested from substances like sugar. The energy is partly converted into forms that the molecular machines called enzymes can use to perform the body's chemistry. Human societies have long depended on stores of chemical energy to meet its needs: wood, coal, oil, and gas, from which energy can be released by burning. Chemical energy is the currency of life. Plants use sunlight to generate and store it, and each biochemical reaction is an energetic transaction. Yet chemical energy is a complex thing. It is often said to be stored in the bonds that link atoms into molecules, but it's not so easy to say how the making and breaking of bonds, involving rearrangements of electrons, translates into energy changes. Chemical reactions may do work by changes in pressure and volume as well as molecules' internal energy, and the direction of chemical change is determined by the so-called free energy, which also involves changes in entropy. So the energetics of chemical change demand the subtle accountancy of thermodynamic theory. And the stability of energy-rich substances is also a matter of reaction speed: Coal stores energy for eons, while dynamite is apt to release it all too readily.

**RELATED TOPICS**
See also
ATP
page 58

OXIDATION
page 100

BURNING
page 102

**3-SECOND BIOGRAPHIES**

ALFRED NOBEL
1833–96
Swedish chemist whose invention of dynamite and gelignite created the wealth behind the Nobel Prizes

JOSIAH WILLARD GIBBS
1839–1903
American scientist who laid the foundations for modern thermodynamics

**EXPERT**
Philip Ball

***Chemical energy is released in a vast number of reactions, from the burning of fossil fuels such as gas, to the energy plants use to grow.***

# NUCLEAR ENERGY

**3-SECOND THRASH**
The source of nuclear energy is the binding energy from the strong nuclear force that holds together the nucleus of an atom.

**3-MINUTE THOUGHT**
Mass contains a lot of energy and nuclear energy illustrates just how much. Particles in the nucleus of atoms surrender tiny amounts of their mass associated with the strong nuclear force—but this results in the release of huge amounts of energy, provided by vast numbers of nuclear reactions. Nuclear fusion powers the Sun while nuclear fission warms the Earth's interior.

Nuclear energy powers the Sun and the stars and warms the Earth's interior. Nuclear energy is released from the nuclei at the heart of atoms. Nuclei are composed of protons and neutrons—except for the hydrogen nucleus, which is one lone proton. All the other elements have more than one proton and these positively charged particles repel each other. This electrical repulsion is overcome by the strong nuclear force—one of the fundamental forces of nature—that binds together the protons and neutrons. The field created by the strong nuclear force makes up much of the mass of protons and neutrons. This mass can also be thought of as energy and it is sometimes called the binding energy of the nucleus. The amount of binding energy needed depends on the size of the nucleus. Elements with bigger, heavier nuclei up to iron need progressively less energy to bind each proton and neutron into the nucleus. The nuclei of atoms heavier than iron need progressively more energy to bind each proton and neutron into the nucleus. Combining the nuclei of lighter elements (nuclear fusion) and splitting the nuclei of heavier elements (nuclear fission) both result in the release of some of this binding energy.

**RELATED TOPICS**
See also
INSIDE A STAR
page 42

FISSION
page 64

FUSION
page 66

**3-SECOND BIOGRAPHIES**
ERNEST RUTHERFORD
1871–1937
New Zealand-born physicist, who discovered the proton and realized that some powerful force must glue these positively charged particles together in the atomic nucleus

ENRICO FERMI
1901–54
Italian-American physicist who supervised the construction of the world's first experimental nuclear fission reactor in Chicago in 1942

**EXPERT**
Leon Clifford

***Fermi and Rutherford were key figures in discoveries involving the nuclear energy that powers the stars and heats the Earth's core.***

# MASS ENERGY

**3-SECOND THRASH**
Mass and energy are two sides of the same coin, as Einstein showed in his famous $E=mc^2$ equation.

**3-MINUTE THOUGHT**
Light is electromagnetic radiation but it also has momentum. So light falling on an object exerts a tiny pressure. We see this effect in our solar system: radiation from the Sun gently sweeps dust particles out into deep space and aligns the dust tails of comets to point away from the Sun. Light pressure could one day be used to power spacecraft equipped with giant solar sails.

Mass and energy are one and the same thing. Mass can be converted into energy and vice versa. Mass can be thought of as the amount of matter within a body and energy as the capacity of a physical system to do work. At first sight they seem very different concepts, and that is what scientists thought at the end of the nineteenth century when they treated mass and energy as separate entities, each subject to its own conservation law. But the realization, following the emergence of James Clerk Maxwell's equations of electromagnetism, that electric fields possess momentum hinted that energy has some kind of mass associated with it. Hendrik Lorentz's theoretical idea that mass increases as bodies approach the speed of light questioned the law of conservation of mass. Then in 1905, German-born physicist Albert Einstein published his special theory of relativity, which showed that mass and energy can be traded with each other according to the relationship since made famous by the equation $E=mc^2$ where $E$ is energy, $m$ is mass, and $c$ represents the speed of light. Because the square of the speed of light is so large, the implication of this equation is that a tiny amount of mass can release huge amounts of energy. This is the secret of nuclear weapons and the reason stars shine.

**RELATED TOPICS**
See also
WHAT IS ENERGY?
page 16

CONSERVATION OF ENERGY
page 32

INFLATION
page 40

**3-SECOND BIOGRAPHIES**
JAMES CLERK MAXWELL
1831–79
Scottish physicist who demonstrated the connection between magnetism and electricity

ALBERT EINSTEIN
1879–1955
German-born physicist who first demonstrated the mathematical relationship between energy and mass made famous by the equation $E=mc^2$

**EXPERT**
Leon Clifford

***As Einstein predicted, in nuclear reactions and the interaction of matter and antimatter, mass and energy are interchanged.***

e+
γ
e+
e-
e-
t

# CONSERVATION OF ENERGY

**3-SECOND THRASH**
You cannot make energy and you cannot get rid of energy; you can only change it from one form to another.

**3-MINUTE THOUGHT**
Conservation of energy is one of a number of fundamental conservation laws in physics. Momentum is also conserved, which is why billiard balls bounce off one another. Angular momentum is another quantity that is conserved and is the reason skaters spin faster when they pull in their arms. These are all manifestations of deeper mathematical relationships involving the existence of symmetries in the equations describing reality.

Energy is conserved. It can neither be created nor destroyed—merely changed into different forms. This is a fundamental law of physics. Since mass is equivalent to energy, we can either think about mass as a form of energy or we can rephrase this law in terms of the conservation of mass-energy; they amount to the same thing. The law of conservation of energy states that the total amount of energy (including mass) in a closed or isolated system is conserved over time—that is, it remains constant. Thus the energy contained in the fuel burned by the internal combustion engine of a car is transformed into the kinetic energy of the car's motion, the heat the engine radiates, the friction of the tires on the road, the noise the vehicle generates, and the movement of the exhaust gases, as well as the vibration in the chassis and other emissions of energy. If we carefully measure all these different forms of energy emanating from a car, we will see that the total exactly matches the amount of chemical energy released from the fuel—none of it disappears and no additional energy mysteriously appears.

**RELATED TOPICS**
See also
WHAT IS ENERGY?
page 16

MASS ENERGY
page 30

INFLATION
page 40

**3-SECOND BIOGRAPHIES**
WILLIAM RANKINE
**1820–72**
Scottish engineer who developed the idea that energy is conserved even when it is transformed—for example, between potential energy and kinetic energy

EMMY NOETHER
**1882–1935**
German mathematician who proved that the conservation laws emerge from symmetry

**EXPERT**
Leon Clifford

***In a pendulum, energy is converted between potential and kinetic energy, also generating heat. Overall the energy is conserved.***

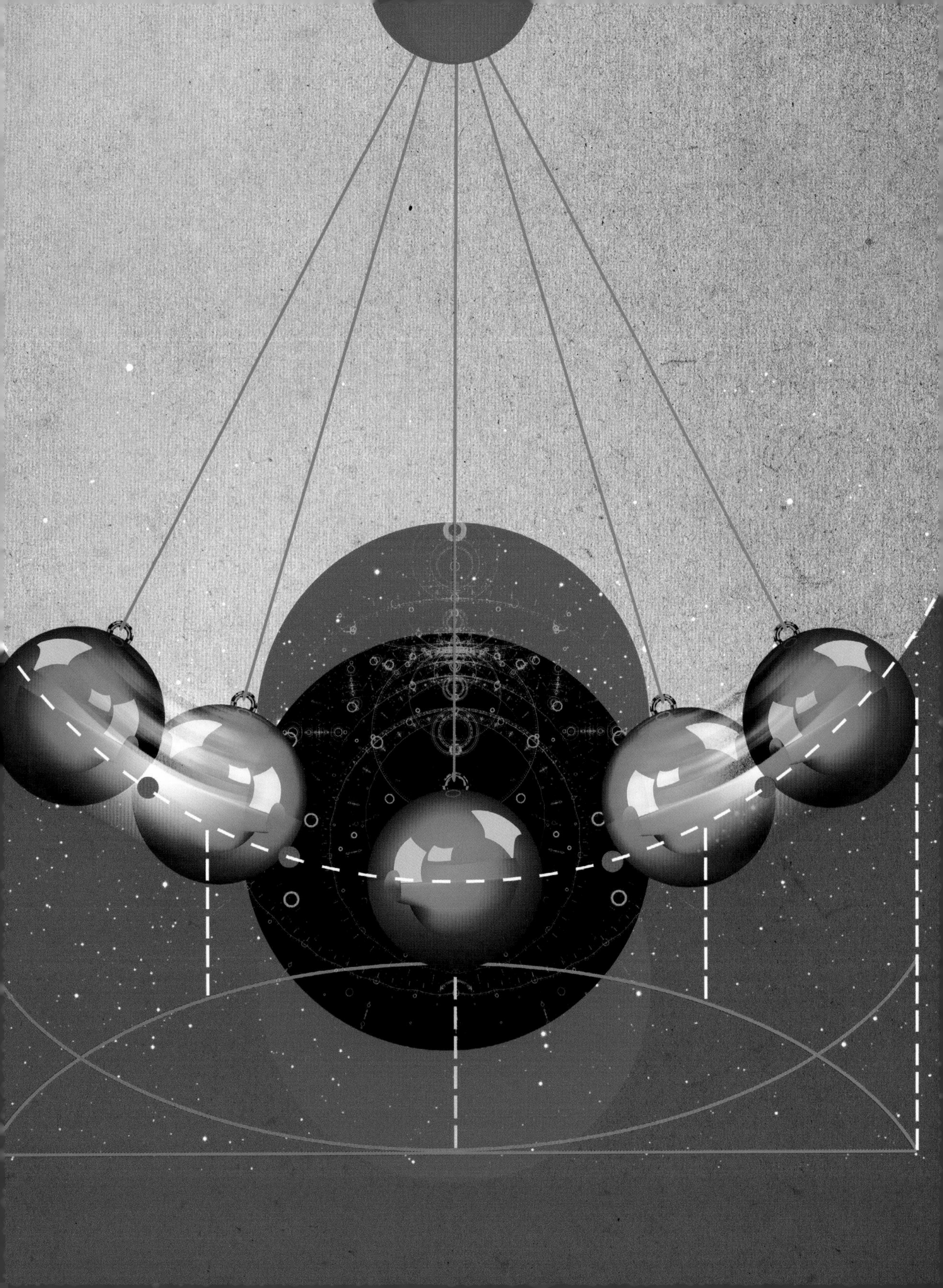

# NATURAL ENERGY

## NATURAL ENERGY
GLOSSARY

**antigravity** Unlike electromagnetism, gravity is always attractive. Antigravity describes a hypothetical ability to nullify gravity or repel matter gravitationally.

**big bang** The best accepted theory of the origin of the universe is the big bang—the universe began as an infinitely dense point of energy that expanded. The term was coined by astronomer Fred Hoyle. Big bang is often used to describe the origin of the universe, but strictly it is the point in time at which the universe began to expand.

**chlorophyll** Several green pigments found in plant cells, which absorb light to enable photosynthesis.

**electromagnetism** During the nineteenth century, it became clear that electricity and magnetism were intimately related; Scottish physicist James Clerk Maxwell showed them to be parts of a single phenomenon: electromagnetism.

**escape velocity** Velocity at which an object traveling away from a massive body will escape that body's gravitational pull without any further force being applied. The escape velocity of the Earth is around 7 miles/sec (11.2 km/sec).

**fundamental field** Physicists use several models to describe the fundamental building blocks of nature, including waves and particles, but often the most productive approach is the field.

**general relativity** Einstein's general theory of relativity provides a mathematical description of the way that matter interacts with spacetime, causing gravity. Einstein struggled with the mathematics and raced with German mathematician David Hilbert to complete the equations.

**gravity waves** One of the predictions of Einstein's general theory of relativity was that when massive objects move back and forth they should generate waves in the gravitational field. Detection of such waves was announced in 2016.

**Heisenberg's uncertainty principle** A major component of quantum physics, the uncertainty principle says that there are linked properties in nature, such as momentum and position, or energy and time, where the more precisely we know one property of a quantum particle, the less accurately we can know the other.

**Higgs field** In the 1960s it seemed that an additional fundamental field was needed to explain why some particles had mass. Named after one of the theory's developers, Peter Higgs, this is the Higgs field. The discovery of the Higgs boson was strong evidence for the existence of the field.

**nuclear reactions within the Sun** The Sun is powered by nuclear fusion reactions. These are reactions in which atomic nuclei merge to form a heavier atom, releasing energy. The dominant reaction in the Sun is hydrogen nuclei fusing to form helium.

**photosynthesis** Process by which plants convert the energy of sunlight into chemical energy.

**shape/geometry of universe** To develop his theory of general relativity, Einstein made use of the mathematics of curved space. This is difficult to envisage, as it is not a curvature *within* space, but of space itself. In principle, the space of the universe could be curved, but it appears to be flat.

**spacetime** Physicist Hermann Minkowski devised the concept of spacetime, a four-dimensional combination of space and time that is essential to understanding Einstein's special theory of relativity.

**steady state theory** The main opposing theory to the big bang in the 1950s and 1960s, steady state suggested that the universe had no beginning but was constantly expanding with new matter coming into being.

**theory of relativity** Einstein's extension of Newton's laws of motion to include light's constant velocity. The result is that it is not possible to separate space and time, as movement influences the passage of time. Unlike general relativity, the special theory of relativity doesn't include gravity.

**vacuum energy** Because of quantum effects, even empty space has a certain amount of energy. There have been attempts to find a way to harness this "vacuum energy," but as this would require a site with less than the minimum possible energy these efforts are unlikely to succeed.

**weak nuclear force** A fundamental force of nature, along with gravity, electromagnetism and the strong nuclear force. The weak force is involved in the decay of atomic nuclei.

# GRAVITY

**3-SECOND THRASH**
Gravity is a fundamental force that acts on energy, provides a source of energy, and helps to bind the universe together.

**3-MINUTE THOUGHT**
The attractive force of gravity causes masses to accelerate towards each other and acquire kinetic energy. As they move closer, gravitational attraction grows, the acceleration due to gravity increases, and the kinetic energy of the system accumulates. But are we getting something for nothing? Doesn't this contradict the law of energy conservation? Not if we treat the gravitational potential energy as "negative energy" that cancels out the positive kinetic energy of motion.

Gravity is a fundamental force of nature, just as is electromagnetism. Gravity is often described as a force that attracts masses together, but mass and energy are equivalent and so gravity is a force that acts between anything with energy, even if that energy is wrapped up in mass. Gravity is an attractive force that weakens with distance but has an infinite range. Anything with mass—or energy—gives rise to a gravitational field, which is described mathematically in Albert Einstein's general theory of relativity as a curvature in spacetime, the fabric of the universe. Gravitational fields are a source of energy called gravitational potential energy. When objects with mass are pulled toward each other by gravity, the gravitational potential energy is converted into the kinetic energy of movement. We use this practically on Earth in hydroelectricity, where gravitational potential energy is converted into the kinetic energy of falling water, which is in turn converted into electrical energy. In extreme cosmological events, such as the collision of two black holes, gravitational potential energy can be released in the form of gravity waves, which can be thought of as rippling undulations in spacetime. So gravity and energy are intimately related: gravity is a property of energy—it acts on energy—and gravitational fields give rise to energy.

**RELATED TOPICS**
See also
MASS ENERGY
page 30

INFLATION
page 40

INSIDE A STAR
page 42

**3-SECOND BIOGRAPHIES**
ISAAC NEWTON
1643–1727
English physicist who first described gravity in scientific and mathematical terms

DAVID HILBERT
1862–1943
German mathematician who derived the field equations of general relativity, the theory of gravitation, from first principles

**EXPERT**
Leon Clifford

***Einstein explained the gravitational force as a warp in space and time that keeps planets in their orbits and makes the apple fall.***

# INFLATION

Our universe is believed to have undergone an almost instantaneous expansion when it was only a tiny fraction of a second old, perhaps just $10^{-37}$ seconds after it came into existence. In an instant, a tiny subatomic volume of space ballooned into an embryonic universe. Spacetime expanded in volume by at least 10,000 trillion trillion times—a factor of $10^{26}$. This rapid, faster-than-light expansion of the very fabric of space lasted just $10^{-35}$ seconds, which is less time than it takes light to travel across an atomic nucleus. It is what cosmologists call inflation. Inflation explains why different parts of our universe are so similar even though they are separated by vast distances and how the distribution of galaxies that we see in the universe came into being. You and I and all the planets and stars we see around us today are made from matter condensed out of energy created in this brief inflationary moment. And this energy—seemingly created out of nothing—came from the gravitational field that built up an offsetting stock of "negative energy" as the universe inflated. The "negative energy" accumulated in the gravitational field balances the mass and energy found in the universe, preserving the law of conservation of energy. Inflation remains a theory that some scientists question, but it appears consistent with what astronomers observe.

**3-SECOND THRASH**

Cosmic inflation expanded the size of our newly created universe by at least 10,000 trillion trillion times in an instant, causing the big bang.

**3-MINUTE THOUGHT**

Inflation may be eternal. Countless trillions of tiny bubbles of spacetime may be inflating into whole new universes every second and for all of time. If true, then the big bang is not the beginning of our universe but just the end of inflation in our part of a much vaster universe. And so there could be infinite copies of our universe, each with an identical Earth and each with a version of you, reading this.

**RELATED TOPICS**

See also
MASS ENERGY
page 30

CONSERVATION OF ENERGY
page 32

GRAVITY
page 38

**3-SECOND BIOGRAPHIES**

EDWIN HUBBLE
1889–1953
American astronomer who realized that the universe was expanding, which implied that it must have had a beginning

ALAN GUTH
born 1947
American physicist who came up with the idea of inflation in 1980

**EXPERT**

Leon Clifford

***In a tiny fraction of a second, the early universe is thought to have expanded far faster than the speed of light in the process known as inflation.***

# INSIDE A STAR

**3-SECOND THRASH**
Inside every star is a vast natural nuclear fusion reactor that is a source of light and heat.

**3-MINUTE THOUGHT**
The heat generated by fusion prevents a star collapsing in on itself due to gravity. For much of its lifetime, a star is balanced between the opposing pressures of its hot gas plasma pushing outward, and the force of gravity pulling inward. Eventually all the nuclear fuel will run out and then the inexorable force of gravity will determine the fate of the star.

Our Sun is the ultimate source of energy that drives our climate and powers life on Earth. And, like all stars, its light and warmth comes from the continual conversion of mass into energy deep within its core through the process of nuclear fusion. Nuclear fusion converts hydrogen into helium and other heavier elements, releasing vast amounts of energy. Stars are massive, with powerful gravitational fields. The mass of a star is pulled inward by this gravity, creating enormous pressure at its core. This pressure overcomes the repulsive electrical force between protons—tiny particles that are the atomic nuclei of hydrogen atoms. Protons are squeezed together close enough for the strong nuclear force—one of the fundamental forces of nature—to take over and bind them together. A series of nuclear reactions results in the formation of helium nuclei. Each nucleus consists of two protons and two neutrons. The field associated with the strong nuclear force provides the source of the energy released during this process. This field makes up most of the mass of protons and neutrons and when they bind together during fusion they give up some of this mass, which is released as energy.

**RELATED TOPICS**
See also
NUCLEAR ENERGY
page 28

MASS ENERGY
page 30

FUSION
page 66

**3-SECOND BIOGRAPHIES**
JEAN PERRIN
1870–1942
French physicist who first proposed that solar energy came from nuclear reactions involving hydrogen

FRED HOYLE
1915–2001
English astronomer who showed how the nuclear reactions in stars produced elements heavier than hydrogen

**EXPERT**
Leon Clifford

***In a star, hydrogen nuclei are fused by the immense temperature and pressure in a multistage process producing helium, the next heaviest element.***

# LIVING THINGS

**3-SECOND THRASH**
Plants take in energy from the Sun through photosynthesis and provide a source of energy for other organisms when they act as food.

**3-MINUTE THOUGHT**
Although photosynthesis appears similar to the action of a photoelectric cell, converting light energy into a usable form, it is far more complex, with chemical processes that can be blisteringly fast, including the quickest chemical reactions on record. Energy from light is captured by special pigments, most frequently the familiar green chlorophyll, and transferred in chemical form to the photosynthetic reaction center, where the storage reaction produces the oxygen we breathe as a by-product.

On a planet teeming with life, there is no getting away from the energy present in living things as they move, interact, and breed. We often describe an individual as being "full of energy" and at one time life was thought to be dependent on a specific type of energy called *vis essentialis* or the "vital spark." This concept is still taken seriously in some Eastern traditions, but science has shown that the energy of life is not an entirely new form of energy, but rather chemical and potential energy put to use by the complex mechanisms of a living organism. The initial source of the energy in almost all living things remains the Sun. Plants make use of this energy directly through photosynthesis, where the energy in photons of light is converted to a chemical form, which can then be used in the survival, growth, and reproduction of the plant. Organisms that don't take energy directly from the Sun usually make use of this stored chemical energy by consuming organic matter, although some take their energy by other means, for example, from the geothermal energy of vents deep under the oceans.

**RELATED TOPICS**
See also
ATP
page 58

OXIDATION
page 100

BIOFUELS
page 120

**3-SECOND BIOGRAPHIES**
JAN INGENHOUSZ
1730–99
Dutch scientist who demonstrated that light was necessary for photosynthesis

MELVIN CALVIN
1911–97
American biochemist who, with Andrew Benson and James Bassham, identified the light-driven reactions behind photosynthesis, known as the Calvin cycle

**EXPERT**
Brian Clegg

***The complex food chains of living organisms are all based on transferring chemical energy, mostly originally produced by photosynthesis from the Sun.***

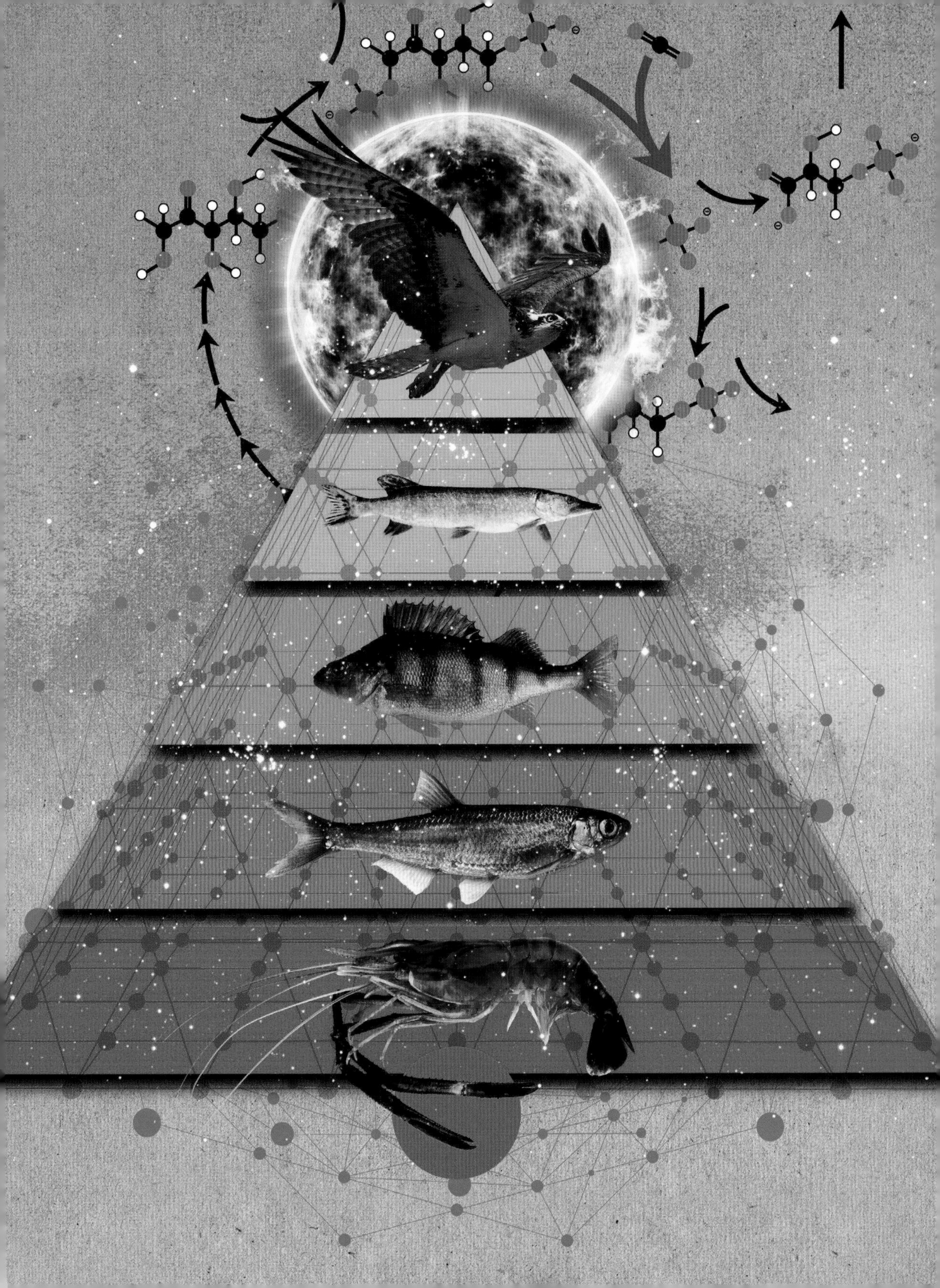

# DARK ENERGY

**3-SECOND THRASH**
Dark energy acts like a kind of antigravity, accelerating the expansion of the universe. It is believed to make up most of the mass of universe—but no one knows exactly what it is.

**3-MINUTE THOUGHT**
Everything we see in our universe is made from what we might call "normal" matter and "normal" energy, the stuff that forms the basis of our physics. Yet this is believed to comprise just 5 percent of the mass of the universe. In recent years, physicists have realized that there is, literally, more to the universe than meets the eye. What else is out there?

Dark energy is believed to make up almost 70 percent of the mass of our universe but scientists do not know exactly what it is. They know that it—or at least something—exists because the expansion of our universe is accelerating rather than slowing down, as you would expect it to do under the influence of gravity. If our theory of gravity is correct, some force must be at work counteracting gravity. Dark energy is the name given to the source of that mysterious force. The acceleration of the universe's expansion has been confirmed by two sets of careful measurements, one looking at faraway exploding stars and one studying the distribution of thousands of galaxies. Furthermore, other measurements confirm that the geometry or shape of our universe is what scientists call "flat," which tells them how much mass is distributed throughout the cosmos. The trouble is we can only find around 30 percent of what theoretically should be there. In other words, most of the mass of the universe is missing. The mass associated with the amount of dark energy needed to account for the observed acceleration of the expansion of the universe, matches up with the amount needed to plug the missing mass gap predicted by the geometry. The existence of dark energy with such an effect is consistent with the description of gravitation contained in Albert Einstein's general theory of relativity.

**RELATED TOPICS**
See also
GRAVITY
page 38

INFLATION
page 40

THE LIFE CYCLE OF THE UNIVERSE
page 152

**3-SECOND BIOGRAPHIES**
SAUL PERLMUTTER, ADAM RIESS & BRIAN SCHMIDT
**born 1959, born 1969 & born 1967**
American astrophysicists who discovered that the expansion of our universe is accelerating, from studying supernova

ALBERT EINSTEIN
**1879–1955**
German-born scientist who developed our best theory of gravity: general relativity, which predicts the existence of dark energy

**EXPERT**
Leon Clifford

***Space is expanding, but this expansion itself is accelerating due to the mysterious dark energy.***

**1947**
Born in New Brunswick, New Jersey

**1968**
Graduates from Massachusetts Institute of Technology (MIT), where he later receives his doctorate

**1971**
Moves to Princeton University, where he works on particle physics

**1971**
Marries Susan Tisch

**1974**
Moves to Columbia University

**1977**
Moves to Cornell University, where he does his first work on the subject of inflation

**1978**
Becomes aware of some of the problems with the big bang theory

**1979**
Begins to work on a possible solution based on inflation

**1979**
Moves to the Stanford Linear Accelerator Laboratory

**1980**
Submits first article on the concept of cosmic inflation

**1980**
Moves back to MIT

**2009**
Awarded the Isaac Newton Medal by the Institute of Physics

**2012**
Awarded the Fundamental Physics prize

# ALAN GUTH

In modern physics, collaboration is the name of the game. But there is still room for individual ideas, and Alan Harvey Guth was responsible for a key contribution to modern cosmology—rescuing the big bang and extending our thinking about energy and the universe.

Guth was brought up in New Jersey, where his father owned a small store. He picked up an interest in science at school, and cites TV shows and books like *The Universe and Dr. Einstein* by Lincoln Barnett as his inspiration. He studied physics at the Massachusetts Institute of Technology (MIT), where he was "fascinated by the idea that the world could be described by precise mathematical laws, so I chose physics because it was the branch of science most closely connected with the quest to discover the fundamental laws." This was a difficult time for physics postdocs, and Guth had a wide range of short posts before he settled into a long-term position.

In the 1960s, the big bang theory had seen off its main competitor—steady state. However, by the 1970s, the big bang had its own issues. Critics argued that the universe was too uniform for the big bang theory to be correct—widely separate areas would not have had the opportunity to come into equilibrium. When Guth first developed an idea that might solve this issue, it was effectively a spare-time activity—he was working in particle physics at Cornell University. But it was after he moved to Stanford that he formally published his concept of cosmic inflation.

His idea of a sudden, incredibly fast expansion of the early universe that switched on and then off again acted as a patch to the big bang theory that made it match observation. To make it work, Guth employed vacuum energy, and required an extra fundamental field, which Guth hoped would be the Higgs field, though there is no evidence to suggest this. After his early success, Guth went on to cover a wide range of concepts in theoretical physics. He had always been interested in magnetic monopoles—particles with a single magnetic "charge" where all the ones we have come across have both positive and negative poles—and continued to work on these, alongside concepts that included the theoretical time machines predicted by general relativity. Inevitably inflation has been a major part of his work. At the time of writing, Alan Guth is still working, having returned to his alma mater, MIT.

*Brian Clegg*

# ZERO-POINT ENERGY

**3-SECOND THRASH**
Space is not as empty as it seems for, according to quantum mechanics, its vacuum is full of energy.

**3-MINUTE THOUGHT**
Quantum mechanics, our theory of the very small, is weird. Not only does it predict the existence of zero-point energy but this means that particles can flick into existence apparently out of nothing. They get their mass by borrowing it from the zero-point energy that fills the vacuum of empty space. Indeed, the subatomic world is believed to be a seething, foaming mass of particles popping into and out of existence.

Quantum mechanics, our theory of the very small, predicts that the vacuum of empty space is filled with energy. The existence of this naturally occurring energy, known as zero-point energy, arises as a result of Heisenberg's uncertainty principle, which shows that at the quantum level pairs of values such as time and energy cannot be known exactly. Zero-point energy is the absolute minimum amount of energy remaining in a quantum mechanical system after all other energy has been removed and temperature has been reduced to the absolute zero point. The uncertainty principle predicts a certain fuzziness to the quantum world that enables tiny fluctuations to occur in the fields that fill space, such as those due to electromagnetism and the strong and weak nuclear forces. It is these inescapable oscillations that give rise to this energy. Quantum mechanics predicts that all space is filled with these oscillations so that although the energy involved at any point is small, the total energy is very large. Indeed, the density of this zero-point energy has been estimated to be vastly greater than the energy density due to nuclear fusion at the Sun's core. The trouble is, none of this energy can be used—because any effort to extract zero-point energy would reduce the energy of the vacuum to below the minimum required by the uncertainty principle.

**RELATED TOPICS**
See also
NUCLEAR ENERGY
page 28

MASS ENERGY
page 30

DARK ENERGY
page 46

**3-SECOND BIOGRAPHIES**
WERNER HEISENBERG
1901–76
German theoretical physicist who developed the uncertainty principle, which predicts the existence of zero-point energy

HENDRIK CASIMIR
1909–2000
Dutch physicist who gave his name to the Casimir effect, a phenomenon believed to be related to the existence of zero-point energy

**EXPERT**
Leon Clifford

***Quantum mechanics allows very short timescale bursts of energy, which provide an unusable foam of background energy throughout the universe.***

# WHERE DID IT ALL COME FROM?

**3-SECOND THRASH**
Although it seems impossible that all the energy in the universe came from "nothing," in principle, physics suggests that this is a feasible option.

**3-MINUTE THOUGHT**
Some theories of what happened "before the big bang" do not present the same problem of understanding as the conventional big bang theory, as they posit the existence of plenty of energy and matter before the big bang and don't require the confusing concept of the energy in the universe emerging from nowhere. An example is the ekpyrotic theory, where our current universe was formed as a result of a collision of two pre-existing "branes" in a larger multidimensional meta-universe.

Energy and matter are the stuff of the universe. The best current theory for the origins of the universe tells us that it expanded in a process known as the big bang, but does not explain *how* the universe came into being in the first place. Given that this process is unknown, we can't assume that conservation of energy applies. It can seem remarkable that all the energy (including all the matter in the universe) we see today came from very little. This is sometimes explained by saying that gravity provides negative energy. Imagine a high-speed object shooting away from the Earth at escape velocity. Gravity slows it to a stop, but it doesn't return. Now it has no kinetic or potential energy. Conservation of energy tells us it must, therefore, have had no energy on the surface of the Earth—so its high kinetic energy must have been countered by a high negative potential energy from the Earth's gravitational field. A physicist will tell you that this picture is flawed, because general relativity means that gravitational energy is not an absolute. General relativity gives a more complex mathematical view—but still means that energy conservation is not an issue for the universe.

**RELATED TOPICS**
See also
WHAT IS ENERGY?
page 16

CONSERVATION OF ENERGY
page 32

GRAVITY
page 38

**3-SECOND BIOGRAPHIES**
ALBERT EINSTEIN
**1879–1955**
German physicist whose general theory of relativity drives the big bang theory

GEORGES LEMAÎTRE
**1894–1966**
Belgian physicist who proposed the expansion of the universe that led to the big bang theory

FRED HOYLE
**1915–2001**
English astronomer who coined the term big bang

**EXPERT**
Brian Clegg

***Through inflation and into the formation of early matter and, eventually, galaxies, the universe evolved from the big bang.***

# STORING ENERGY

**base** In genetics, a base (more strictly a nucleobase) is one of the five compounds found in pairs in the "tread" part of the spiral staircase shape of DNA, or providing links between sections of the simpler structure of the related molecule RNA. The bases are adenine, cytosine, guanine, thymine (in DNA only), and uracil (in RNA only).

**electrochemical cell** Device that uses chemical energy to generate electricity by freeing up electrons to flow through a conductor. Common examples of electrochemical cells are batteries and fuel cells.

**electrode/electrolyte** An electrode is a conductor, such as a piece of metal or carbon, that is given an electrical charge in a cell or other device. An electrolyte is a substance, usually a fluid or gel, containing ions. The ions conduct electricity by moving through the electrolyte, attracted to electrodes that dip into it.

**heavy elements** The weight of an atom is primarily determined by the number of particles—protons and neutrons—that make up its central nucleus. Elements from hydrogen to iron are considered light, while those with more particles in the nucleus are called heavy.

**hydroelectricity** The generation of electricity as a result of the movement of water, typically by building a dam and allowing the flow of water as it drops downhill to run through a turbine, turning its blades to produce electricity. The process converts potential energy via kinetic energy to mechanical energy, and finally electrical energy.

**ions** Atoms that have an overall positive charge due to losing one or more electrons, or an overall negative charge as a result of gaining electrons.

**isotopes** The chemical properties of an element are determined by the number of electrons it has, which is identical to the number of protons in the nucleus. However, atoms of the same element can have differing numbers of neutrons in the nucleus. Atoms with the same number of protons but different numbers of neutrons, such as uranium-235 (143 neutrons) and uranium-238 (146 neutrons), are called isotopes.

**lead-acid battery** The first type of battery that was capable of being charged multiple times, the lead-acid battery (or more accurately the lead-acid cell) consists of two lead electrodes dipped into a sulfuric acid electrolyte. When the battery is charged, the outside of one electrode is oxidized to lead oxide, while during discharge the electrodes both become covered in lead sulfate.

**lithium-ion battery** Rechargeable batteries most commonly used in electronic devices such as cell phones. Such cells typically consist of a carbon electrode and a lithium alloy oxide electrode in an electrolyte of lithium salts.

**phosphate group** A group of atoms with a single atom of the element phosphorous linked to four atoms of oxygen. Phosphate groups are notably found in adenosine triphosphate (ATP), the compound used by living organisms to store energy in chemical form.

**respiration** At the most basic level, respiration is the mechanism by which a living cell converts chemical nutrients into energy. At a higher level, the term is often used to describe the mechanism used to carry oxygen and carbon dioxide around a multicellular organism.

**sugar** An organic molecule such as glucose, fructose, and sucrose, one of a family of sweet-tasting water-soluble carbohydrates (molecules containing carbon, hydrogen, and oxygen).

**uranium-235/uranium-238** The two most significant isotopes of uranium used in nuclear reactors and atomic bombs. The numbers 235 and 238 refer to the total number of protons and neutrons. Each has the same number of protons—92—however uranium-238 has three more neutrons than uranium-235.

# ATP

**3-SECOND THRASH**
ATP is the constantly recycled currency of energy that is ultimately used to pay for every living process.

**3-MINUTE THOUGHT**
If ATP is the currency of energy, then ATP synthase is its minting machine. Housed in the mitochondria, protons produced by respiration power this enzyme to crank out ATP from inputted ADP. Given its ubiquity, ATP synthase is believed to have formed early in the history of life on Earth. It's considered an example of modular evolution—that is, a composite of functionally independent subunits that combined to form a structure with a different ability.

From bacteria to human beings, adenosine triphosphate (ATP) is the currency of energy in all living cells. Every process that goes on in your body—such as muscle contraction and the making of new proteins—requires energy sourced from ATP. ATP is a comparatively small, soluble molecule and so is able to diffuse quickly and supply energy wherever it's needed. It's made up of three distinct parts: A base (adenine), a sugar (ribose), and three phosphate groups linked together like a tail. These phosphate groups have the formula ($PO_4^{3-}$)—a phosphorus atom bonded to four oxygen atoms. A large amount of energy is released when the relatively unstable bond between the last phosphate group and the one before is broken, resulting in adenosine diphosphate (ADP) and phosphate (typically denoted by $P_i$). It has been estimated that we "consume" our body weight in ATP every single day, which is all the more amazing when you consider that at any time we contain only about 2 ounces (60 g) of the stuff. This is possible because ADP and $P_i$ can be recombined to form ATP again if the necessary energy is supplied. Both plants and animals use glucose to remake ATP through the process of respiration. This takes place in the mitochondria, the so-called powerhouses of the cell. A single glucose molecule can be used to produce up to 38 ATP molecules.

**RELATED TOPICS**
See also
CHEMICAL ENERGY
page 26

CONSERVATION OF ENERGY
page 32

**3-SECOND BIOGRAPHY**
FRITZ LIPMANN
1899–1986
German-American biochemist who identified ATP as the main source of energy in the cell in his 1941 paper "The Metabolic Generation and Utilization of Phosphate Bond Energy"

**EXPERT**
Simon Flynn

***A bond in the relatively simple adenosine triphosphate (ATP) molecule is used to transport chemical energy around living organisms.***

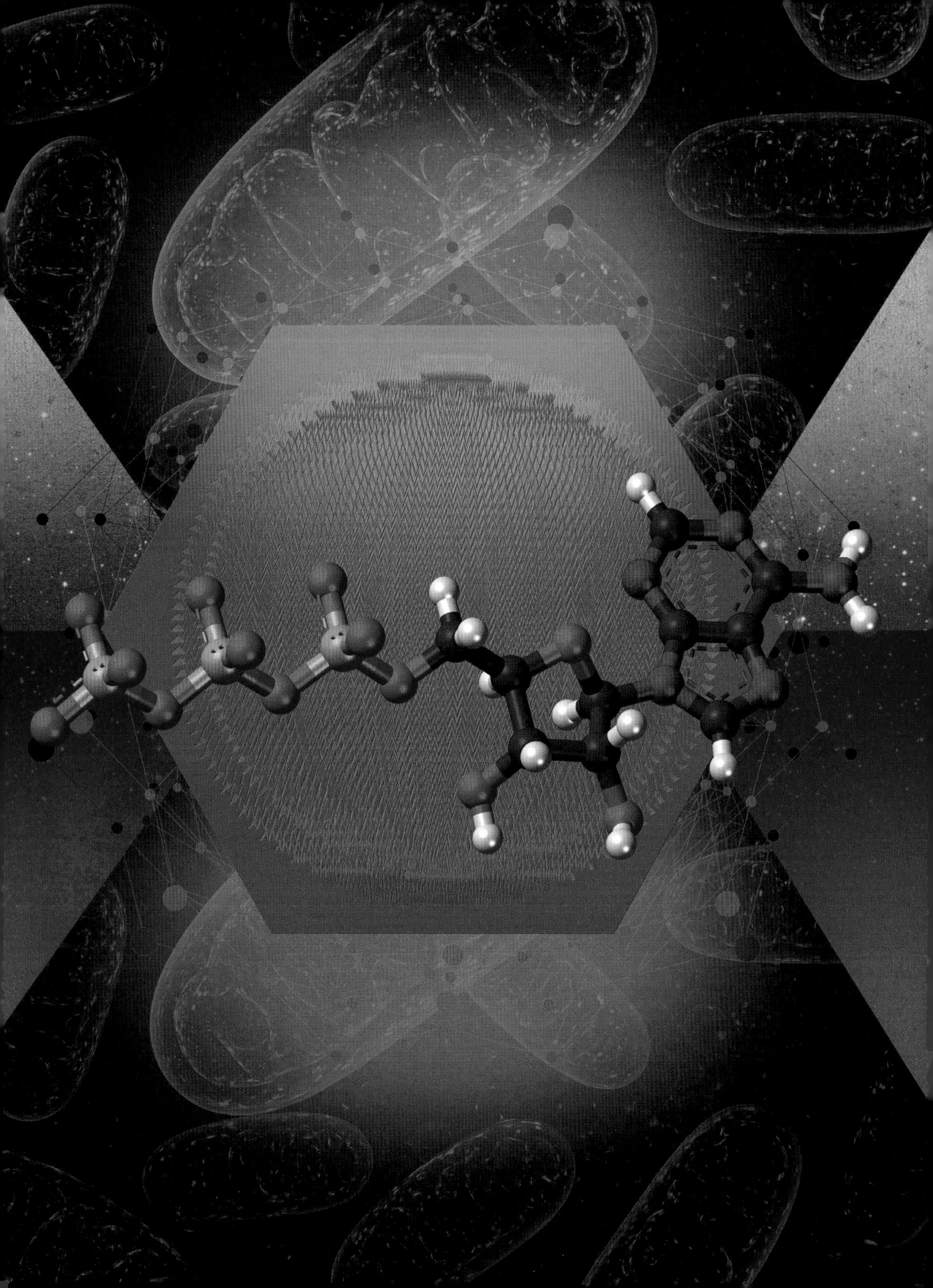

# COAL

**3-SECOND THRASH**
Coal is made of plant material that originally lived in swampy or boggy conditions 100–300 million years ago, containing a high percentage of carbon.

**3-MINUTE THOUGHT**
Use of coal is unlikely to reduce in the near future. Known reserves are expected to last at least 100 years, it produces energy more cheaply than oil and natural gas, and vast amounts are found in the major economic powers of the United States, China, and India. For many nations, it provides realistic energy security, given that oil and gas are mostly found in the Persian Gulf. But it's a dirty fuel, particularly in its mining and burning.

Coal is an energy-dense fossil fuel. It currently provides almost 30 percent of all humanity's energy needs and is responsible for the production of more than 40 percent of the planet's electricity. Around 300 million years ago, there were many more swamps and bogs on the Earth's surface. Typically, when plant material dies, organisms break it down and, through the process of respiration, return much of the carbon contained in living matter back to the atmosphere in the form of carbon dioxide ($CO_2$). However, swamp water contains low concentrations of oxygen and this limits decomposition. Dead plants sink to the bottom and get covered by increasing amounts of silt. This process keeps repeating and pressure increases. Eventually, water is squeezed out of the plant cells and peat is formed. The peat is buried yet farther, pressure and temperature increase and, in order of increasing carbon concentration, the following types of coal are formed: lignite, bituminous, subbituminous, and anthracite. Bituminous is the most commonly used: 60–80 percent carbon, it also contains sulfur impurities. As a result, sulfur dioxide ($SO_2$) as well as $CO_2$ is formed when bituminous coal is burned—$SO_2$ is a major contributor to acid rain and increasing concentrations of $CO_2$ in the atmosphere are lowering the oceans' pH.

**RELATED TOPICS**

See also
CHEMICAL ENERGY
page 26

OIL
page 62

BIOFUELS
page 120


**3-SECOND BIOGRAPHIES**

JAMES WATT
1736–1819
Scottish businessman and inventor whose steam engines drove the Industrial Revolution and were powered by coal

HUMPHRY DAVY
1778–1829
English scientist whose safety lamp, designed to avoid igniting flammable gases, saved the lives of many coal miners

**EXPERT**
Simon Flynn

***Over time, dead organic matter is put under pressure by a buildup of soil above, gradually forming a layer of coal.***

# OIL

Crude oil is a mixture of hydrocarbons (molecules that contain only hydrogen and carbon), varying in size from the smallest, methane ($CH_4$), to much larger molecules containing tens of carbon atoms. The hydrocarbons can be separated from the mixture using fractional distillation, in which crude oil is vaporized and then separated off in fractions due to the molecules' differing boiling (or condensation) points. These fractions include: refinery gas, gasoline, naphtha, kerosene, lubricating oil, and bitumen (various tars). Crude oil is formed in a very similar way to coal. Organisms (mostly plankton) living in ancient oceans died and fell to the seabed. Silt and sand covered them and this sediment steadily increased over time. Pressure and temperature increased and sedimentary rocks such as sandstone, limestone, and shale were formed. The organic matter was gradually broken down into simpler and simpler molecules, resulting in underground reservoirs of oil and natural gas. These reservoirs, buried deep underground and difficult to access, are actually porous rock with the oil contained in tiny spaces. As a result, it typically requires considerable effort to extract oil. Once refined, it's most commonly used to produce the energy required to make things move such as cars, boats, and planes.

**3-SECOND THRASH**

Crude oil is a mixture of hydrocarbons—molecules made of carbon and hydrogen only; these were originally made up of ocean-based organisms such as plankton.

**3-MINUTE THOUGHT**

In the United States, about 47 percent of refined oil is used as gasoline, almost 10 percent as jet fuel, and 30 percent as diesel and other fuels. The hydrocarbons that make up crude oil aren't just a useful energy resource. They are also used to produce lubricants, plastics such as polythene and PVC, solvents such as ethanol, and fabrics such as nylon. Current proved reserves of oil should last 50 years.

**RELATED TOPICS**

See also
CHEMICAL ENERGY
page 26

COAL
page 60

BIOFUELS
page 120

**3-SECOND BIOGRAPHIES**

HIERONYMUS BRUNSCHWIG
c. 1450–1512
German surgeon and alchemist who wrote *Liber de arte distillandi*, the first book on distillation

RUDOLF DIESEL
1858–1913
German inventor of the internal combustion engine

EDWARD BUTLER
1862–1940
English inventor credited with making the first gasoline engine

**EXPERT**

Simon Flynn

***Oil is a mix of hydrocarbons formed from decayed organic matter in deep underground reservoirs.***

# FISSION

Nuclear fission occurs when the nucleus of a heavy element, containing many protons and neutrons, splits into two or more smaller nuclei. It is typically triggered when the parent nucleus absorbs an extra neutron. As well as releasing a large amount of energy, the fission reaction produces additional free neutrons, which can go on to trigger the same reaction in neighboring nuclei. As long as there is a "critical mass" of fissionable material, this gives rise to a self-sustaining chain reaction. If such a reaction proceeds very rapidly, the result is a nuclear bomb—the most notorious application of fission. However, the reaction can be tamed by employing just a small percentage of fissionable nuclei, such as the isotope uranium-235, embedded in a more stable isotope like uranium-238. Further fine-tuning can be achieved through the use of control rods of a different material, which absorb free neutrons without fissioning. Reactors built on these principles produce energy in the form of heat, which can then be converted to rotary motion using a steam turbine. In a nuclear submarine, the turbine directly drives the propulsion system, while in a nuclear power station it is used to run an electrical generator.

**3-SECOND THRASH**
Nuclear fission produces large amounts of energy by splitting heavy atomic nuclei into smaller ones; fission reactors are used in nuclear submarines as well as power stations.

**3-MINUTE THOUGHT**
The end-products of nuclear fission are typically highly radioactive; in other words, their nuclei are unstable and prone to decay with the release of radiation that is hazardous to life. A nuclear explosion may scatter radioactive fallout over a wide area, and even the carefully managed disposal of radioactive waste from a nuclear reactor poses significant environmental concerns. Many of the strongest arguments against both nuclear weapons and nuclear power generation center on the potentially catastrophic effects of radioactivity.

**RELATED TOPICS**
See also
NUCLEAR ENERGY
page 28

FUSION
page 66

NUCLEAR
page 134

**3-SECOND BIOGRAPHIES**
LISE MEITNER
1878–1968
Austrian-Swedish physicist who produced the first theoretical analysis of nuclear fission and the energy it releases

OTTO HAHN
1879–1968
German scientist who worked with Meitner and made the first experimental discovery of fission

**EXPERT**
Andrew May

***When a nucleus splits in a chain reaction it generates a number of neutrons, each of which can cause another nucleus to split.***

# FUSION

### 3-SECOND THRASH

Nuclear fusion powers the Sun and offers a tantalizingly attractive energy source for the future—but only after major engineering challenges have been overcome.

### 3-MINUTE THOUGHT

The first generation of nuclear weapons, such as the Hiroshima and Nagasaki bombs of August 1945, were fission devices. Later thermonuclear weapons, such as the so-called "hydrogen bomb," are two-stage devices using both fission and fusion. The heat generated by an initial fission reaction triggers fusion in the second stage, and high-energy neutrons produced by this fusion reaction enormously boost the rate of fission. Nevertheless, most of the explosive energy of the bomb still comes from fission, not fusion.

The vast quantities of energy radiated by the Sun come predominantly from the fusion of hydrogen nuclei—each comprising just a single proton—into helium nuclei, each made up of two protons and two neutrons. Fusion is a nuclear reaction just like fission, but it has several important advantages. It generates more energy per unit mass, its waste products are less hazardous, and the basic fuel—hydrogen—is far more abundant than fissionable elements such as uranium. Unfortunately, reproducing starlike nuclear fusion in a power station poses huge engineering challenges. Fusion can only occur at extremely high temperatures, measured in tens of millions of degrees, at which hydrogen forms an ionized gas called a plasma. The problem is not just to create the plasma, but to keep it contained once it has been created. The Sun's plasma is held in place by its gravity, but on Earth alternative methods must be used. Chief among these are magnetic confinement devices such as the tokamak, originally invented as long ago as the 1950s. Limited, laboratory-scale fusion was achieved soon after, but subsequent progress has been painfully slow. Even today, the energy produced by experimental fusion reactors is outweighed by the energy needed to generate and confine the plasma.

### RELATED TOPICS

See also
NUCLEAR ENERGY
page 28

INSIDE A STAR
page 42

FISSION
page 64

### 3-SECOND BIOGRAPHIES

ARTHUR EDDINGTON
**1882–1944**
English astrophysicist who first suggested that nuclear fusion takes place inside stars

ANDREI SAKHAROV
**1921–89**
Russian nuclear physicist who did important work on both the hydrogen bomb and the tokamak

### EXPERT

Andrew May

***One of the most promising nuclear fusion reactors, the tokamak, uses a doughnut-shaped vessel in which high-temperature plasma is contained by a magnetic field.***

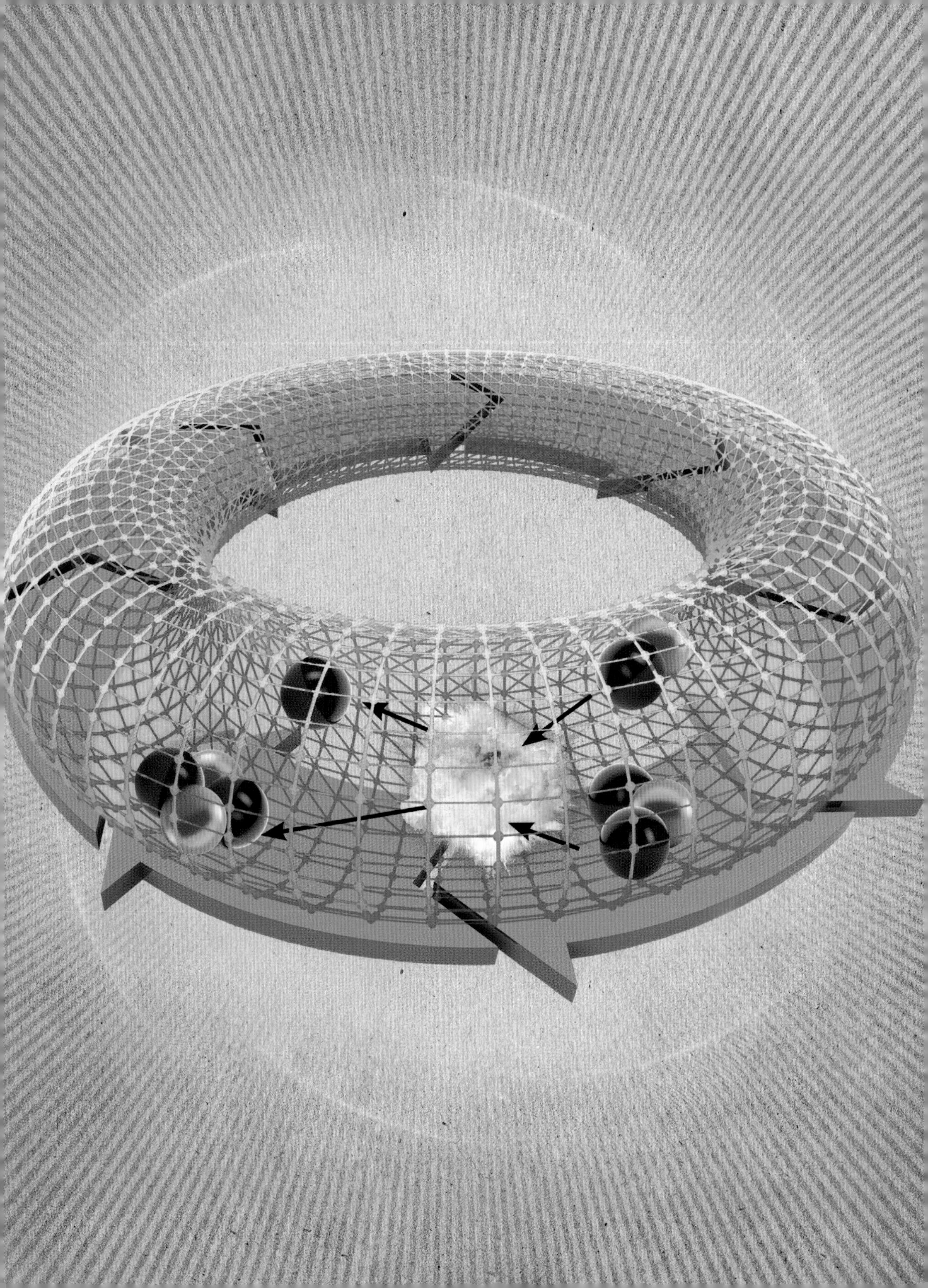

**1745**
Born in Como, then part of the Duchy of Milan

**1769**
Writes a treatise, *On the Forces of Attraction of Electric Fire*

**1774**
Becomes an instructor at Como grammar school

**1775**
Invents a "perpetual electrophotus"

**1778**
Is the first person to isolate pure methane

**1779**
Becomes professor of experimental physics at the University of Pavia

**1785**
Is made rector of the University of Pavia

**1794**
Is the first foreigner to win the Royal Society's Copley Medal; marries Teresa Peregrini

**1798**
Luigi Galvani dies

**1800**
Discovery of the "electric pile" announced by the Royal Society

**1801**
Receives a gold medal and pension from Napoleon Bonaparte

**1810**
Is made a count by Napoleon

**1827**
Dies at home in Camnago

# ALESSANDRO VOLTA

Alessandro Volta was born the same year that saw the invention of the Leyden jar, the first means of storing electrical charge and, arguably, the most influential electrical instrument of the eighteenth century. The Leyden jar ushered in a golden age of research into electricity and Volta proved to be one of the most important protagonists in developing the subsequent revolution of electrochemistry.

Volta's greatest discovery, the voltaic pile or battery, was prompted by the work of fellow countryman Luigi Galvani. An anatomist by background, Galvani had spent several years dissecting frogs when his research interests took a dramatic turn in 1780. One such dissection was lying on a franklin square, a device similar to a Leyden jar, when the inner nerve of its leg was touched by Galvani's metal scalpel and violent contractions were suddenly observed. Similar results were seen when the steel scalpel touched a brass hook that was being used to hold a leg in place. This, along with a series of follow-up experiments, led Galvani to confirm and extend the contemporary theory of animal electricity and 1791 saw the publication of his *Commentary on the Effects of Electricity on Muscular Motion*. Only 12 copies of the first edition were printed and Galvani sent one of these to Volta.

At first, Volta was a supporter of Galvani's conclusions. A change came when Volta started to look carefully at his compatriot's experiments, repeating many of them himself. Thus began the controversy that divided the scientific community. Where Galvani saw the generation of the electric current as occurring in the frog's muscles, Volta's brilliance was to realize that it was the presence of two different metals that caused the current. It was Volta's belief that the frog was essentially acting as a very sensitive electrometer and he began to investigate how the quantity of electricity produced by two different metals could be increased.

This led to his invention of the electric pile, what we would today call a battery. After experimenting with various pairs of metals, Volta settled on zinc and silver as providing the greatest effect. He also recognized that Galvani's frogs had provided something essential to the process: a conductive liquid. So, Volta's first electric pile consisted of alternating zinc and silver disks, with brine-soaked cardboard between each disk and wires connected at both ends. All other things being equal, the size of the voltage and the current produced depended on the metals used and could always be increased by adding more disks. Volta named this new apparatus the "artificial electric organ" in reference to the torpedo fish, or electric ray.

This was the first time a steady current could be produced and, in the years following Volta's discovery, versions of his pile were used to extract a number of elements such as sodium and calcium, thereby leading to their discovery, as well as to separate water into hydrogen and oxygen.

*Simon Flynn*

# WATER STORAGE

**3-SECOND THRASH**
Pumped storage is a way to store energy resources for hydroelectricity by using off-peak power to pump water back up into a reservoir.

**3-MINUTE THOUGHT**
Pumped storage can use seawater. Japan has a 30-megawatt seawater plant at Okinawa, built in 1999, which pumps water from the shoreline to an artificial reservoir 500 feet (150 m) high. If pumping in such plants is conducted at high tide and released at low tide, there is an extra gain in energy. Some is extracted from the tide itself. Other seawater pumped-storage projects are being explored in Hawaii, Chile, Ireland, and the Middle East.

One big challenge for energy generation is that demand fluctuates. There are peak times—in the mornings, say, when people are boiling kettles. But demand plummets in the middle of the night. Yet energy generation can't always be cranked up or down accordingly—nuclear power stations, for example, have to run more or less flat out. So electricity grids need "load balancing," using a process by which unwanted energy—which is relatively cheap—can be stored for use at peak times. Massive batteries are an expensive solution, but for hydroelectric energy there is a better plan. The water itself can be pumped back up the mountainside to the high reservoir using off-peak power, and then discharged during peak demand. Such pumped-storage hydroelectricity has the largest storage capacity for any form of grid energy, since it is much easier to store water than electricity itself. Some of these storage plants are more or less closed systems: The same water is discharged and then pumped back between two reservoirs. Because of inevitable energy losses in generation and pumping, the process consumes energy overall—typical efficiencies are around 70–80 percent. These plants can still be economically viable, however, because the pumping power is low-cost, and they can also act as short-term reserves, for example to compensate for breakdowns of conventional power stations.

**RELATED TOPICS**
See also
POTENTIAL ENERGY
page 22

HYDRO
page 126

**3-SECOND BIOGRAPHIES**
WILLIAM RANKINE
1820–72
Scottish engineer who coined the term "potential energy"

EDWARD MACCOLL
1882–1951
Scottish engineer and pioneer of hydroelectricity, who conceived one of the first pumped-storage plants at Cruachan on Loch Awe in the 1930s

**EXPERT**
Philip Ball

***In a pumped storage system, water is pumped up to a high mountain reservoir, and stored as potential energy until there is a demand.***

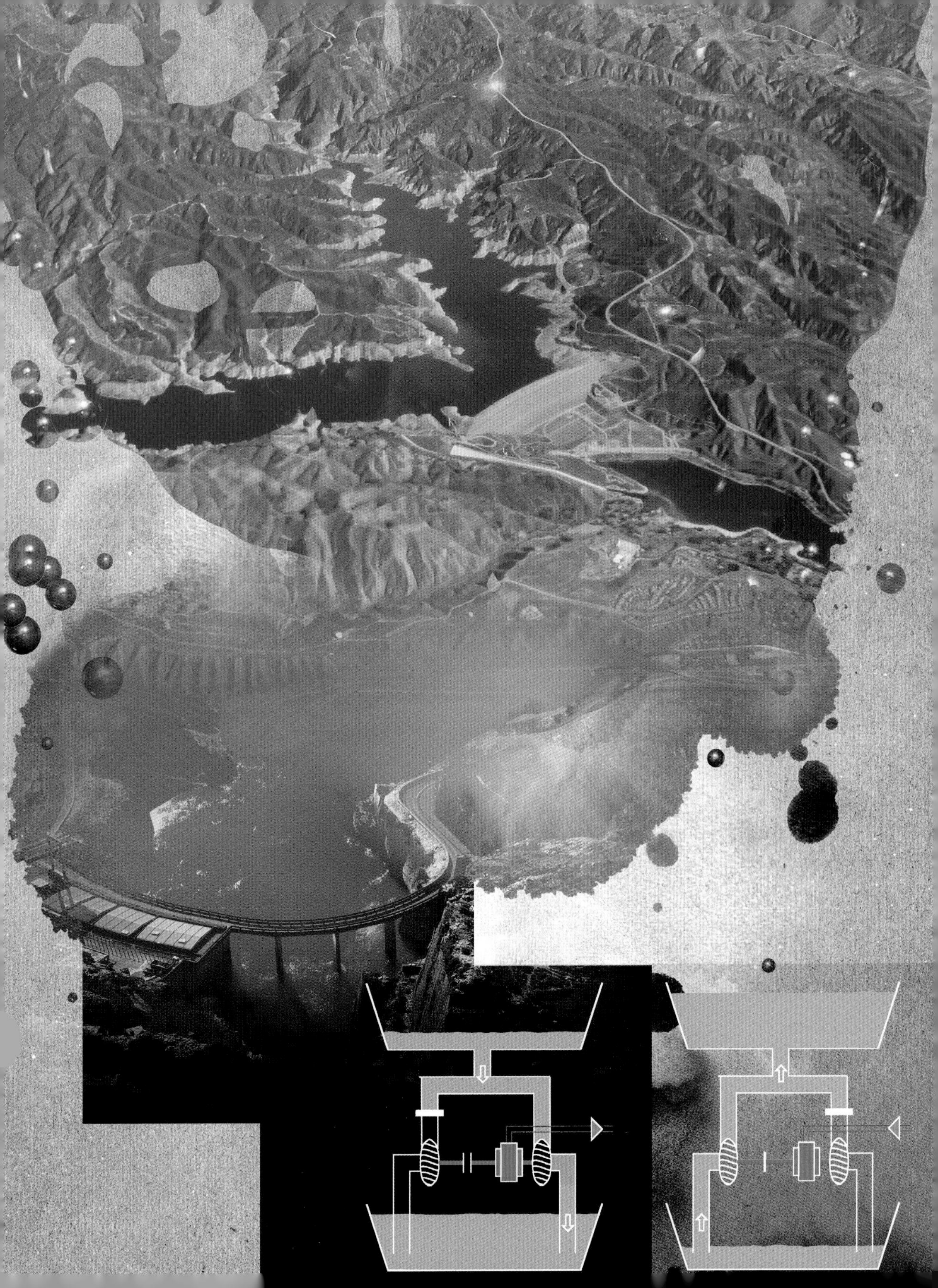

# FUEL CELLS

**3-SECOND THRASH**
Fuel cells harness the chemical energy of combustion by turning it directly into electricity in a battery-like cell fed with fuel.

**3-MINUTE THOUGHT**
As pretty much any chemical reaction involves movement of electricity, all manner of chemical processes can be harnessed in fuel cells. "Microbial fuel cells" exploit the biochemical metabolic processes of microorganisms to generate (generally small amounts of) electricity from decomposition reactions in soils, sediments, and wastewater. Other fuel cells are coupled to cell reactions in higher living organisms. One runs on digestion of human blood sugar by encapsulated yeast, generating electricity that might power biomedical implants.

The burning of fuels in oxygen is basically a transaction in electrons. The fuel loses them, the oxygen gets them, and atoms are rearranged in the process. But it's a rather wasteful way to generate electricity—as is done in coal- or gas-fired power stations, for example. In turning the chemical energy first to heat and then to electricity, a fair amount of it is squandered. It would be more efficient to use the chemical energy directly to set electrons in motion, making electricity without the intermediation of heat. That's what fuel cells do. Invented in the 1830s, they conduct the combustion process in two separate halves of an electrochemical cell. At the positive electrode, electrons are extracted from the fuel, which is typically hydrogen gas, methane, or methanol. The electrons flow along a wire to the negative electrode, where they are released onto oxygen gas. The two half-cell reactions thus produce the products of the combustion reaction: For a hydrogen-powered fuel cell, these are positive hydrogen ions and negative hydroxide ions, which combine to make water. The two half-cells are connected by an electrolyte, generally an ion-transporting liquid or solid. Modern fuel cells are compact, mostly solid devices that act as fuel-powered batteries for vehicles, portable electronics, spacecraft, and satellites and medium-scale power generation.

**RELATED TOPICS**
See also
BATTERIES
page 74

OXIDATION
page 100

BURNING
page 102

**3-SECOND BIOGRAPHIES**
WILLIAM GROVE
**1811–96**
Welsh scientist and barrister, generally credited as the inventor of the fuel cell, which he first reported in 1838

FRANCIS THOMAS BACON
**1904–92**
English engineer (and descendant of his seventeenth-century namesake) who developed the first hydrogen fuel cell in the 1940s and 1950s

**EXPERT**
Philip Ball

***Grove's concept of the fuel cell was devised long before it could be practically used, but provides an efficient way of using hydrogen to generate electricity.***

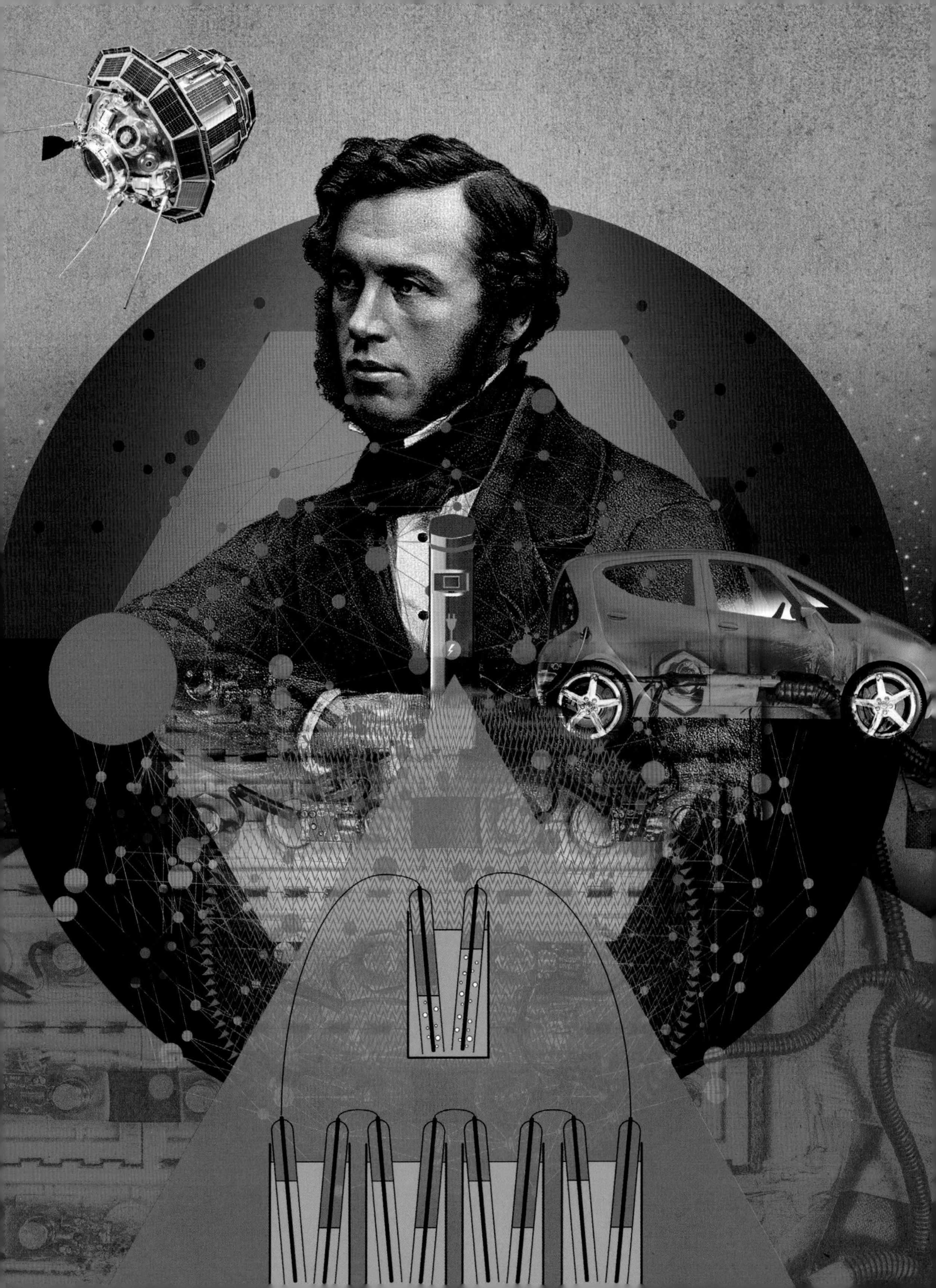

# BATTERIES

Electrical batteries have been around for more than 200 years, but they are as important today as ever. Rapidly developing technologies such as cell phones and electric vehicles would be impossible without the compact, self-contained energy source that is so easy to take for granted. In essence, a battery is a device for converting stored chemical energy into electrical energy. The simplest battery—a single cell—consists of two electrodes of different materials embedded in an electrolyte of a third material. When the electrodes are connected to an external circuit, an electric current flows through it as a result of chemical reactions occurring inside the battery. Some types of battery, called primary batteries, are ready to deliver power as soon as they are constructed, but once they are fully discharged they cannot be recharged. The cylindrical alkaline batteries commonly used in torches and TV remote controls fall in this category. In contrast, secondary batteries—such as the lead-acid battery found in a car, or the lithium-ion batteries used in cell phones and laptop computers—must be charged from an external power supply before they can be used, but can then be discharged and recharged over and over again.

**3-SECOND THRASH**

Batteries store energy in chemical form, and deliver it as electricity; depending on the specific design they may be either single-use or rechargeable.

**3-MINUTE THOUGHT**

In some applications, notably emergency systems or military weapons such as missiles and torpedoes, a battery may go unused for years—or even decades—before suddenly needing to operate with high reliability. Such situations require a special type of battery, called a reserve battery, in which the electrolyte is only activated immediately before use. For example, the electrolyte may be inert in its normal solid form, but becomes active when heat is applied to liquefy it.

**RELATED TOPICS**

See also
CHEMICAL ENERGY
page 26

ALESSANDRO VOLTA
page 68

FUEL CELLS
page 72

**3-SECOND BIOGRAPHIES**

ALESSANDRO VOLTA
1745–1827
Italian electrical experimenter who constructed the world's first battery—the electric pile

GASTON PLANTÉ
1834–89
French physicist who invented the rechargeable lead-acid battery

**EXPERT**

Andrew May

***Batteries have come a long way from Volta's electric pile, and are now capable of storing sufficient energy to power a sports car.***

# TRANSMITTING ENERGY

**TRANSMITTING ENERGY**
GLOSSARY

**absolute zero** The lowest conceivable temperature: -459.67°F (-273.15°C). This temperature is not achievable in practice, because it implies that all particles are at their lowest energy and not moving. This is not possible for quantum particles, as Heisenberg's uncertainty principle does not allow them to have exactly known position and momentum.

**BCS theory** Bardeen Cooper Schrieffer theory, which explains the functioning of low-temperature superconductors.

**convector** Substance such as a gas or liquid that can transfer heat by convection.

**Cooper pairs** Pairs of electrons produced when a substance is at a very low temperature. They can move through the substance without any electrical resistance.

**electrical conductors** Substances such as metals or carbon that have an abundance of free electrons; these electrons enable the conductor to carry an electrical current.

**electromagnetic spectrum** Visible light is part of a much wider range of electromagnetic radiation, going from radio waves, through microwaves, infrared, visible, ultraviolet, and X-rays to gamma rays.

**free electrons** Electrons that are not bound to an atom and can move through a material carrying electrical current or heat.

**greenhouse effect** Some molecules in the atmosphere, such as carbon dioxide and water vapor, absorb infrared light and re-emit it. When they absorb infrared emitted by the Earth's surface and send some of it back toward Earth, they create a "greenhouse effect" that retains heat. We need the effect—without it the average temperature would be -0.4°F (-18°C)—but too strong a greenhouse effect produces excessive surface temperatures.

**Heisenberg's uncertainty principle** A major component of quantum physics, the uncertainty principle says that there are linked properties in nature, such as momentum and position, or energy and time, where the more precisely we know one property of a quantum particle, the less accurately we can know the other.

**ITER experimental fusion energy reactor** International project to develop a reactor in France using a fusion reaction similar to that of the Sun to generate energy. It is expected to go live in the 2020s.

**laser** Device using stimulated emission of radiation to produce a beam of visible light. The name stands for "light amplification by the stimulated emission of radiation."

**maglev trains** Trains that use very powerful magnets to repel the track so they float above it. They require superconducting magnets.

**maser** Device using stimulated emission of radiation to produce a beam of microwaves. The name stands for "microwave amplification by the stimulated emission of radiation."

**MRI scanners** Medical scanners that use powerful magnetic fields to turn water molecules in the body into tiny radio transmitters, building up an internal picture of the body.

**plasma** Fourth state of matter after solids, liquids, and gases. A plasma is gaslike, but is made up of ions rather than atoms and is an electrical conductor.

**quantized states** Possible states a particle or other quantum object can be in. These states cannot have any value, but have fixed values, requiring a jump between states.

**quantum mechanics** Old term for quantum physics—the physics of the very small, such as electrons, atoms, and photons of light.

**speed of light** Light travels at a constant speed in any medium. It is fastest in a vacuum, where its speed of 299,792,458 m/sec (983,571,056 ft/sec) is the highest possible velocity through space.

**stimulated emission** Emission of photons of light by an electron, triggered by an incoming photon. One photon comes in and two emerge, so the light is amplified.

**superconductivity** Quantum effect when a material is extremely cold, which means that there is no resistance to electrical currents. As a result, currents can travel indefinitely without heat losses, and very high currents can be produced that are used to provide extremely powerful magnets.

**vacuum** Region of space containing hardly any matter.

# THERMAL CONDUCTION

**3-SECOND THRASH**
Heat is conducted through solids by atomic vibrations and by free electrons, which are abundant in good electrical conductors—making them good thermal conductors, too.

**3-MINUTE THOUGHT**
Conduction helps transfer heat from one fluid (liquid or gas) to another inside a heat exchanger. Heat exchangers save on energy consumption by using waste heat—such as that generated by machines or furnaces—to heat buildings or vehicles. Inside the simplest designs, one fluid flows through a pipe while the other fluid passes around the outside of that pipe. While heat transfers within each fluid via convection, conduction transfers heat through the pipe wall.

We feel the effects of thermal conduction whenever we stir a hot cup of coffee with a metal spoon. Heat is conducted from the parts of the spoon in contact with the hot liquid toward the initially colder end that pokes out from our coffee. Conduction occurs thanks to both the vibration of the spoon's atoms and the movement of electrons inside the metal. Heat causes a solid's atoms to vibrate about their normal positions, and the higher the temperature the more they vibrate. As the atoms in solids are joined by interatomic bonds, the vibrations in hot areas will pass from atom to atom through the solid, transferring the heat. This is a relatively slow process compared with the conduction of heat energy by free electrons. These electrons gain kinetic energy (energy due to movement) from the heat, and when they move to a colder area and collide with an atom, this energy is converted into vibrational energy and the temperature of that area rises. Free electrons move quickly, so this method of transferring heat is faster. Metals are particularly good thermal conductors because they contain lots of free electrons. Aluminum alloys and copper, for example, are used in electronics to conduct heat away from processors and so prevent malfunctioning due to overheating.

**RELATED TOPICS**
See also
HEAT
page 18

KINETIC ENERGY
page 20

DOWN THE WIRE
page 92

**3-SECOND BIOGRAPHIES**
DANIEL GABRIEL FAHRENHEIT
1686–1736
German physicist who made the first mercury thermometer in 1714

ANDERS CELSIUS
1701–44
Swedish astronomer who, in 1742, devised a temperature scale in which 0 degrees was the temperature of boiling water and 100 degrees the temperature of melting ice; these fixed points are swapped over in the Celsius scale we use today

**EXPERT**
Sharon Ann Holgate

***The relatively free electrons in metals make them good thermal conductors to help heat escape from a hot source.***

# CONVECTION

**3-SECOND THRASH**
In convection, energy is carried between locations when fast-moving particles are carried through a fluid, such as water, so that they transfer their energy to a different location.

**3-MINUTE THOUGHT**
Convection often takes place in "cells"—volumes within a fluid that contain rotating material. If the fluid is warmed from below, for example, a portion of it rises, carrying heat upward. The fluid then falls back down as it cools, and this process tends to form alternate clockwise and counterclockwise rotations. A large-scale example of convection cells are the Hadley cells, where air rises near the equator, heads toward the pole, then descends and returns toward the equator.

Convection appears to be the poor cousin of the three basic mechanisms for energy to get from place to place, and yet it is vital for the Earth's systems. In effect, convection is conduction that plays piggyback. Just as in conventional conduction, the kinetic energy of molecules in a source object is transferred to kinetic energy in the molecules of the convector. But instead of passing that energy along in a chain-like fashion, the convector now moves from one place to another, carrying the energy with it, before passing it on. This movement takes place in fluids—liquids, gases, or plasmas—and in its natural form is dependent on basic thermodynamics or diffusion to power the movement. The most dramatic examples of convection on Earth are the weather systems. Here two forces combine—hot fluids rise, as they are less dense because the increased kinetic energy of their molecules make them more dispersed, while interaction between different adjacent zones plus the rotation of the Earth causes sideways motion. With human intervention, the energy for movement can be deliberately inserted, rather than coming from natural circulation—as, for example, when a convector heater uses a fan to blow hot air from place to place.

**RELATED TOPICS**
See also
HEAT
page 18

KINETIC ENERGY
page 20

THERMAL CONDUCTION
page 80

**3-SECOND BIOGRAPHIES**
GEORGE HADLEY
1685–1768
English lawyer and meteorologist who proposed a mechanism for the trade winds now known as Hadley circulation

GILBERT WALKER
1868–1958
English physicist who discovered the weather convection process known as Walker circulation

**EXPERT**
Brian Clegg

***The convection currents carrying heat through water in a kettle and through the atmosphere may be on a different scale but involve similar processes.***

# RADIATION

It's estimated that more energy from the Sun strikes Earth in 90 minutes than was consumed by the world in 2001. This reaches us via radiation, the transfer of energy through electromagnetic waves. These make up the electromagnetic spectrum—in order of increasing energy, radio waves, microwaves, infrared, visible light, ultraviolet, X-rays, and gamma waves. Electromagnetic waves don't need a medium (matter) to transmit and so can travel through space (a vacuum). More than 99 percent of all energy expended on Earth came originally from the Sun. This is because, within the Sun, hydrogen nuclei are fused together to form helium nuclei and energy, which radiates out in all directions. Traveling at the speed of light, this takes almost eight and a half minutes to reach us, mostly as infrared and visible light. All objects radiate energy. The hotter something is the more energy it emits (gives out). At room temperature, most objects emit infrared and this radiation moves into, and through, the visible spectrum as the temperature rises. This explains why a heated iron poker first glows red, then yellow, and then a bluish white as its temperature increases. The color of an object affects its ability to emit and absorb electromagnetic waves. White is a poor absorber and emitter, black is the opposite.

**3-SECOND THRASH**

Radiation is the only energy transfer that can take place in a vacuum. It works through the emission of electromagnetic waves such as infrared and visible light.

**3-MINUTE THOUGHT**

By day, Earth absorbs energy from the Sun and heats up—if we walk on a beach at night the sand feels warm. This absorbed energy is subsequently emitted by Earth in the form of infrared radiation. Many atmospheric molecules such as water, methane, ozone, and carbon dioxide absorb this energy, trapping it. This is the greenhouse effect. Without it, Earth's average temperature would be below water's freezing point. Life on Earth would be impossible.

**RELATED TOPICS**

See also
HEAT
page 18

INSIDE A STAR
page 42

SOLAR
page 122

**3-SECOND BIOGRAPHIES**

WILLIAM HERSCHEL
1738–1822
German-born British musician and astronomer who discovered infrared while investigating the reaction of a thermometer at different positions on a spectrum

JAMES CLERK MAXWELL
1831–79
Scottish physicist who was first to fully explain the nature of electromagnetic radiation

**EXPERT**

Simon Flynn

***Herschel uncovered "invisible light" in the form of infrared, part of the wider spectrum of electromagentic energy reaching Earth from the Sun.***

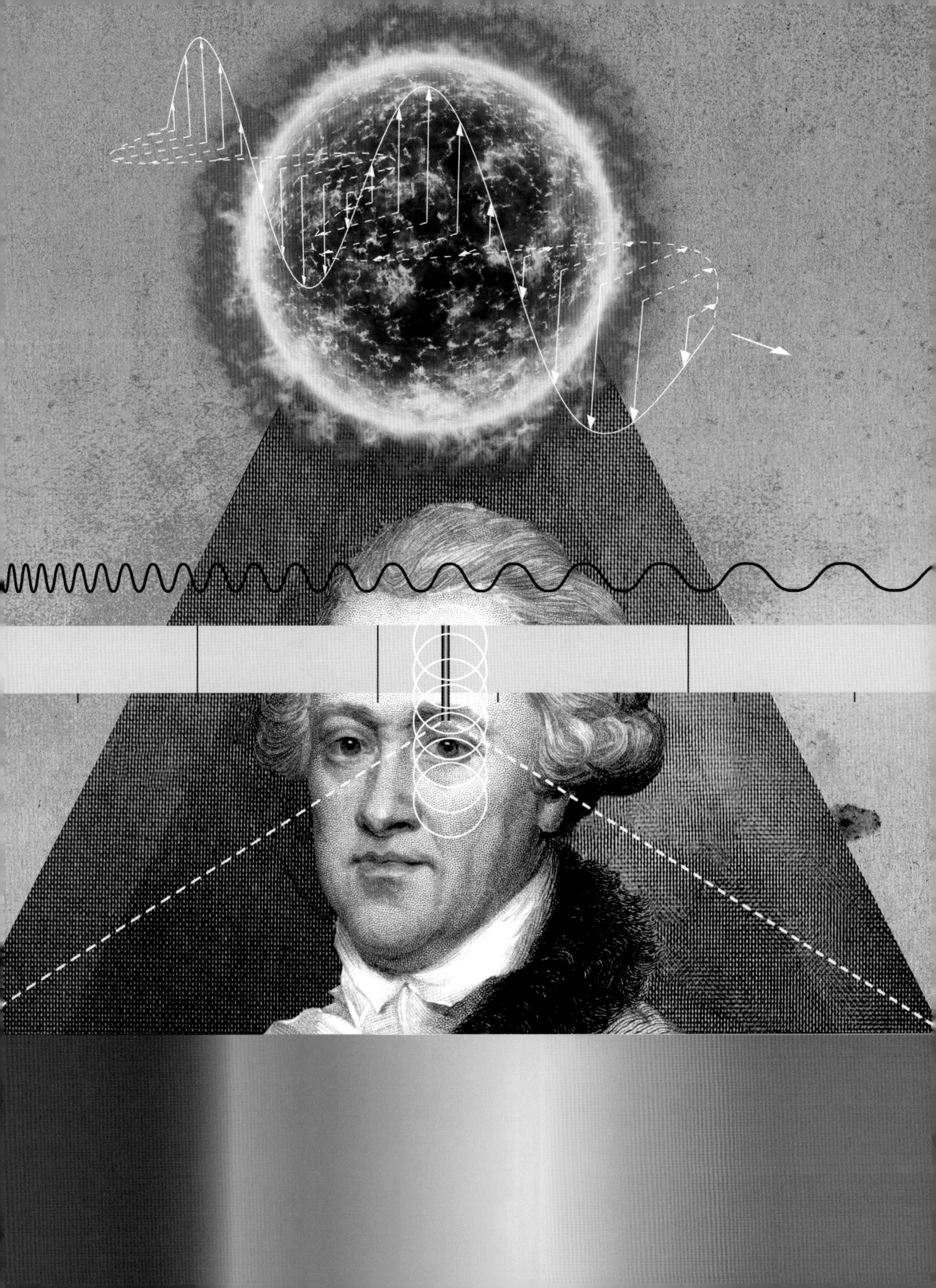

1853
Born in Groningen, Netherlands

1870
Enrolls at the University of Groningen

1871
Moves to Heidelberg

1873
Returns to Groningen

1879
Gains doctorate in physics

1878
Moves to Delft Polytechnic as assistant to the director

1882
Becomes professor of experimental physics at the University of Leiden

1883
Appointed a member of the Royal Academy of Science of Amsterdam

1887
Marries Maria Adriana Wilhelmina Elisabeth Bijleveld

1904
Opens low-temperature laboratory at Leiden

1908
Liquefies helium

1911
Demonstrates superconductivity in a mercury wire

1913
Awarded the Nobel Prize in physics

1926
Dies in Leiden, Netherlands

# HEIKE KAMERLINGH ONNES

Although his name is not one that will be familiar to many, Heike Kamerlingh Onnes was a worthy winner of the 1913 Nobel Prize in physics. His award was for "his investigations on the properties of matter at low temperatures, which led, inter alia, to the production of liquid helium." Kamerlingh Onnes was, without doubt, the early twentieth century's king of cool.

He came from a solid, middle-class family. His father, Harm, owned a brickworks and his mother, Anna, was an architect's daughter. Kamerlingh Onnes studied physics at Groningen before going on to study under Bunsen and Kirchoff at Heidelberg. Subsequently he received his doctorate back in Groningen for work on mechanisms such as Foucault's pendulum that can be used to study the rotation of the Earth. He went on to explore liquids, and by the 1880s, when he moved to Leiden, he began to focus particularly on low temperature work.

Primarily an experimental physicist, Kamerlingh Onnes wanted to demonstrate the theoretical developments of Johannes Diderik van der Waals and Hendrik Lorentz, and after a number of years working on cooling mechanisms, he became the first to successfully liquefy helium in 1908, reaching temperatures close to 1 K (-457.9°F or -272.2°C), just a degree away from the ultimate low-temperature limit of absolute zero. Kamerlingh Onnes would make a range of discoveries on the behavior of matter at extremely low temperatures, but his most remarkable result in 1911 was superconductivity, where he found that materials at extremely low temperatures behaved totally unexpectedly. When he took mercury below 4.2 K (-452.11°F or -268.95°C), its resistance entirely disappeared.

Some had expected that resistance would become infinite at absolute zero. Others, like Kamerlingh Onnes himself, thought that resistance would gradually taper off, but this sudden disappearance was shocking. The total absence of resistance was also difficult to prove experimentally. One approach taken by Kamerlingh Onnes was to start a current flowing in a loop of superconducting material encased in liquid helium. He then monitored the current from outside by measuring the magnetic field it produced, which remained constant. Although Kamerlingh Onnes could only do this for a few hours before his helium boiled away, a variant of the experiment was run in the 1950s for 18 months without any reduction in current flow.

Originally Kamerlingh Onnes referred to the effect as "supraconductivity," though he would later accept the more popular superconductivity. Six years after his death, the building where he worked was renamed the Kamerlingh Onnes laboratory.

***Brian Clegg***

# TRANSPORTING CHEMICALS

**3-SECOND THRASH**
Fossil fuels cram in a lot of energy, which makes them efficient at transmitting energy from place to place, especially with a pipeline network such as natural gas.

**3-MINUTE THOUGHT**
Hydrogen is sometimes cited as an alternative means of powering cars, and it packs in even more energy per unit weight than does an oil-based fuel. The big advantage of hydrogen is that it burns cleanly, giving off no carbon dioxide ($CO_2$), just water. However, it is more dangerous to store and transport than gasoline, and would need a new distribution network. It also takes up to six times as much space as conventional fuel for the same energy, reducing capacity.

We are so used to stopping at a gas station for fuel or to putting wood and coal in a stove that it's easy to forget that these fuels represent a way to transport chemical energy to where it can be transformed into electricity, heat, or mechanical energy. The reason that oil-based fuels such as gasoline and kerosene remain attractive as a way to transmit energy is that they pack in a high energy density. Aviation fuel, for example, holds 15 times as much energy per kilogram as does the explosive TNT. The reason we use TNT for demolition is the speed with which it burns—releasing all its energy very quickly. Kerosene and gasoline burn much slower, but manage to hold a great deal of energy. These oil-based fuels store around 100 times as much energy per kilogram as a good battery, which is why batteries are only just becoming practical for cars and still can't pack in enough power for a plane. Natural gas tends to be lumped in with coal and oil, but has the big advantage that many countries have a pipeline distribution network in place, reducing the costs of transmitting energy in this form.

**RELATED TOPICS**
See also
CHEMICAL ENERGY
page 26

COAL
page 60

OIL
page 62

**3-SECOND BIOGRAPHIES**
NIKOLAUS OTTO
1832–91
German engineer behind the first practical internal combustion engine using an oil-based fuel

KARL BENZ
1844–1929
German engineer who built the first car powered by an internal combustion engine

VLADIMIR SHUKHOV
1853–1939
Russian engineer who devised the first cracking process to turn crude oil into practical fuels

**EXPERT**
Brian Clegg

***Oil-based fuels pack in far more energy per unit weight than equivalents such as batteries.***

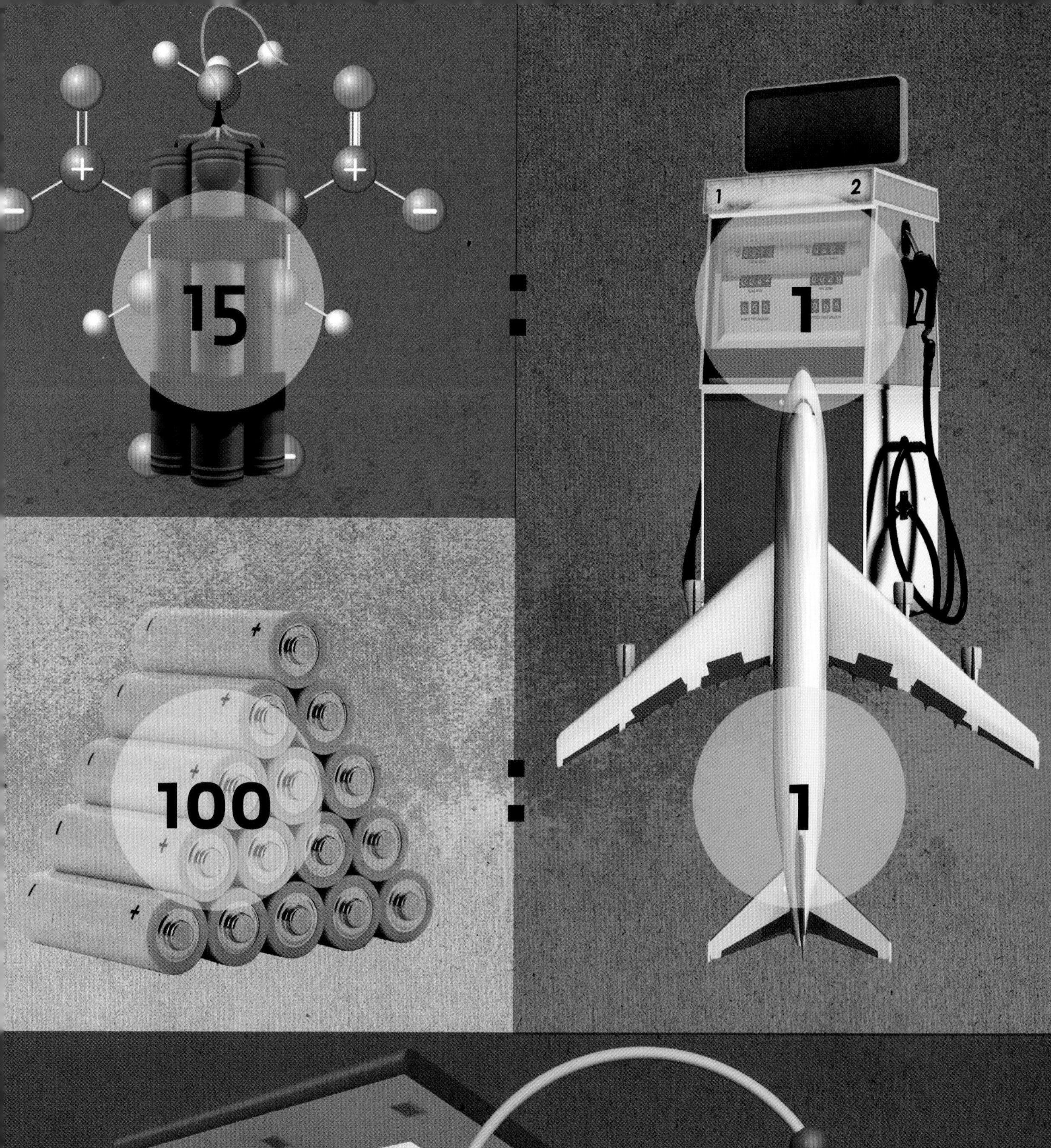
15
1
100
1

# LASERS

### 3-SECOND THRASH
Lasers produce intense, coherent light waves from the way light emission from one atom can stimulate that in another.

### 3-MINUTE THOUGHT
There are natural lasers of a sort in space. Stimulated emission of microwave radiation (maser action) can occur in the molecular clouds from which stars form. This emission has the very narrow wavelength range characteristic of the laser process, and was first detected in 1965 coming from hydroxyl (OH) molecules in such clouds. Astrophysical masers have since been discovered for other molecules, including water and methanol; some are stimulated by starlight.

Energy may be absorbed and emitted by atoms and molecules when they make jumps between the fixed amounts of energy that quantum mechanics limits them to having. Their rotations, vibrations, and electron energies all have such "quantized" states, and a jump between two states may be accompanied by absorption or emission of a photon—a packet of light energy—of the corresponding energy. When an atom loses energy, it releases a photon of very specific wavelength and color. In general, such transitions are independent from one atom to the next—in a piece of hot, glowing metal, each atom emits light without regard to its neighbors. But in 1917, Albert Einstein realized that the emission of a photon from one atom could stimulate that from another. Then the two photons are in step (coherent). The peaks and troughs of the light waves coincide. If such photons could be confined within the emitting material, there could be an avalanche effect that induces all the atoms to emit coherently more or less at once. This "light amplification by stimulated emission of radiation" is the physical basis for a device named after the acronym: laser. Because the light is coherent and emitted all in the same direction, it creates a bright, narrow beam that is not easily scattered.

### RELATED TOPICS
See also
FUSION
page 66

RADIATION
page 84

QED
page 110

### 3-SECOND BIOGRAPHIES
CHARLES H. TOWNES
1915–2015
American physicist who first demonstrated the laser principle in 1953 with a device operating at microwave frequencies

THEODORE MAIMAN
1927–2007
American engineer who made the first working laser at visible-light frequencies, using a ruby crystal, in 1960

### EXPERT
Philip Ball

***Theodore Maiman's first laser started an optical revolution that would transform communications through fiber optics and make optical disks possible.***

# DOWN THE WIRE

**3-SECOND THRASH**
Electrical energy in the form of an electromagnetic wave is transmitted along wires, using high voltages for long distances and usually employing alternating current (AC).

**3-MINUTE THOUGHT**
We have traditionally used high-voltage AC to transmit electrical energy over large distances, but there is growing interest in using high-voltage DC that can reduce both the cost and energy losses. This was not considered practical until the mid-twentieth century, because transformers to change voltage have to be AC and switching high-power transmitted DC to and from AC was not viable until a new generation of conversion technology came onstream.

In the twenty-first century electrical wiring is ubiquitous, from the huge cables slung between pylons, to the power supply that charges a smartphone. That electrical distribution has become so commonplace emphasizes both the significance of electricity to civilization and the ease with which this form of energy can be transmitted from place to place. Early descriptions of electricity described it as a form of fluid, as if we poured a pile of electrons into one end of a wire so that they could flow down like water through a pipe and emerge at the other end. The flow of electricity was thought to be from positive to negative pole, and is still conventionally drawn this way, though we now know that the flow of electrons is in the opposite direction. It is true that electrons move down a cable this way. But they drift along at less than walking pace. When a current "flows" in a wire, an electromagnetic field is set up, transmitted at near light speed, starting electrons moving almost immediately at the far end. We mostly transmit electricity as alternating current (AC), where the direction of flow switches rapidly back and forth, rather than single-direction direct current (DC). Transformers push the AC to high voltages for long-distance transmission, because this reduces the current flow, and hence loss, due to heating.

**RELATED TOPICS**
See also
BATTERIES
page 74

THERMAL CONDUCTION
page 80

SUPERCONDUCTORS
page 94

ELECTROMAGNETISM
page 108

**3-SECOND BIOGRAPHIES**
THOMAS EDISON
1847–1931
American inventor who pioneered electrical distribution and championed the DC system

NIKOLA TESLA
1856–1943
Croatian-born American engineer who contributed significantly to the development of the modern AC system

**EXPERT**
Brian Clegg

***An AC electricity supply can be pushed up to high voltages using transformers, reducing power line losses.***

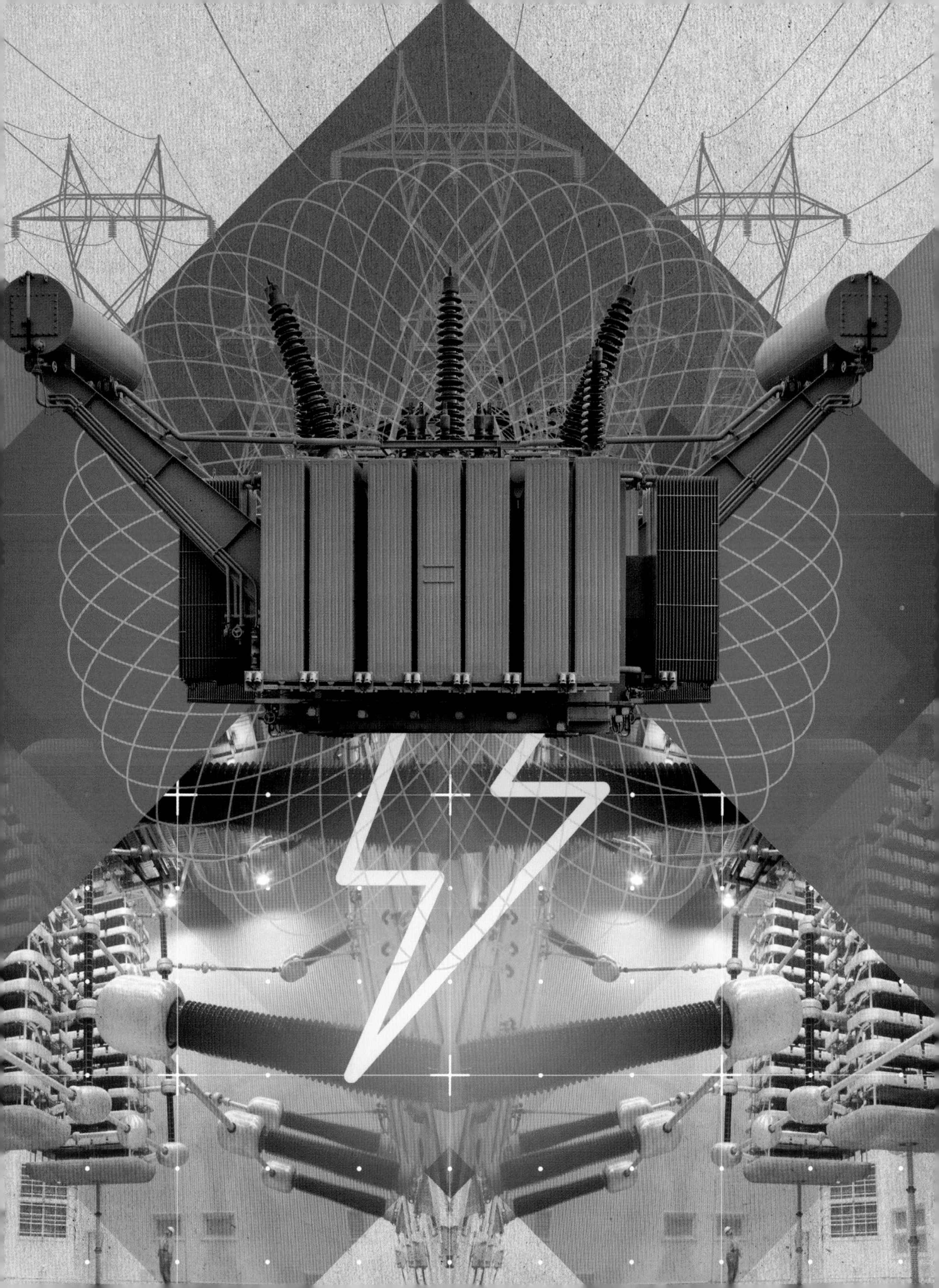

# SUPERCONDUCTORS

**3-SECOND THRASH**
A direct electric current could theoretically flow in a closed loop of superconducting wire forever, as superconductors offer no resistance to the flow of electricity.

**3-MINUTE THOUGHT**
Electromagnets made from superconducting wire generate high magnetic fields, and use little power because an electric current established in the wire flows almost indefinitely. Superconducting electromagnets help levitate and propel maglev trains, and enable MRI scanners to create detailed images of our internal organs and tissues. Ten thousand tons of superconducting magnets will initiate, confine, and shape the plasma inside the ITER experimental fusion energy reactor currently under construction in France.

In 1911, Heike Kamerlingh Onnes found that when solid mercury was cooled below 4.2 K (-452.11°F or -268.95°C) it lost all resistance to the flow of an electric current. The mercury had become a superconductor. Superconductivity has, since then, been shown in thousands of other materials—including compounds containing copper—when they are cooled below each material's specific superconducting transition temperature. This temperature is usually extremely low, but there are "high-temperature" superconductors that show superconductivity at temperatures above 77 K (-321.07°F or -196.15°C). In the 1950s, physicists John Bardeen, Leon Neil Cooper, and John Robert Schrieffer worked together to explain the superconductivity seen at very low temperatures. This resulted in the "BCS theory" (an acronym from their names) in which electrons form into "Cooper pairs" at very low temperatures thanks to a quantum mechanical interaction that creates a slight attraction between the electrons in each pair. While a single electron taking part in an electric current meets with resistance to its movement when it collides with the atoms of the solid it is traveling through, a Cooper pair is only prevented from moving along, and so creating a current, if any collisions with atoms provide enough energy to split the pair up. Since this rarely happens, the Cooper pairs move through the solid with no resistance.

**RELATED TOPICS**
See also
FUSION
page 66

DOWN THE WIRE
page 92

ELECTROMAGNETISM
page 108

**3-SECOND BIOGRAPHIES**
HEIKE KAMERLINGH ONNES
1853–1926
Dutch physicist who discovered superconductivity in 1911

JOHN BARDEEN, LEON COOPER & JOHN SCHRIEFFER
1908–91, born 1930 & born 1931
American physicists who shared the 1972 Nobel Prize in physics for developing the BCS theory of superconductivity

**EXPERT**
Sharon Ann Holgate

***In maglev trains, superconducting magnets are used to suspend the train above the track and to provide the motive power to reach high speeds.***

N
S
N
S
N
S
N
S
N
N
S
S
N
S
N
S
N
S

# CONVERTING ENERGY

## CONVERTING ENERGY
GLOSSARY

**AC** Alternating current (AC), where the voltage in the system varies smoothly between a positive value and the same negative value. Most countries' AC systems go through this cycle 50–60 times a second. AC is used because its voltage can be stepped up or down using a transformer, making it easy to use very high voltages (which lose less power to heat) for long-distance transmission.

**anthropogenic carbon dioxide emissions** The levels of carbon dioxide in Earth's atmosphere have varied significantly over the lifetime of the planet. Levels have increased from around 270 parts per million to more than 400 since the Industrial Revolution. There is very strong evidence that these are human-caused or anthropogenic emissions, as opposed to natural emissions from volcanoes, for example.

**battery** A portable device for storing electrical charge, usually a cell containing a combination of electrodes and an electrolyte. Technically a battery should comprise multiple cells, but the term is now also used for single cells.

**carbon nanotubes** Extremely strong tubes made up of a single layer of carbon atoms.

**Carnot engine** A hypothetical engine that produces work by moving energy from a hot to a cold place.

**electrode/electrolyte** An electrode is a conductor, such as a piece of metal or carbon, that is given an electrical charge in a cell or other device. An electrolyte is a substance, usually a fluid or gel, containing ions. The ions conduct electricity by moving through the electrolyte, attracted to electrodes that dip into it.

**electromagnetic induction** If two electrical wires are near each other and one carries an electrical current of changing voltage, the magnetic field the first wire produces starts a current in the second wire. Producing such a current without direct contact is induction.

**free radical** Chemical substance, usually a molecule, in which an atom has one or more electrons free to join another atom, making the substance very reactive.

**fullerenes** Unusual molecules of carbon that form soccer ball-like spheres or tubes. They are named after the architect Buckminster Fuller, who devised domes with a similar structure.

**greenhouse gas** Gas producing a greenhouse effect, such as carbon dioxide or water vapor.

**quantum theory** Physics theory describing the behavior of the very small particles that make up matter and light.

**respiration** At the most basic level, respiration is the mechanism by which a living cell converts chemical nutrients into energy. At a higher level, the term is often used to describe the mechanism used to carry oxygen and carbon dioxide around a multicellular organism.

**rusting** Chemical process in which iron atoms react with oxygen to produce iron oxide.

**second law of thermodynamics** Sometimes paraphrased as "things that run down," the second law says that in an isolated system, when two bodies are in contact, heat flows from the hotter to the colder. This can be phrased as entropy—the level of disorder in the system—stays the same or increases.

**"step-down" transformer** Device using electromagnetic induction. One coil of wire carries a high-voltage electrical current. This induces a current in a second, nearby coil. If the second coil has fewer turns of wire than the first, the voltage in the second coil is lower, making this a step-down transformer.

**turbofan** Conventional jet engine in which burning fuel expands gases to turn a turbine. This is connected to a fan, which draws air through the engine, supporting the combustion and providing thrust.

**turboprop** Hybrid between a propeller and jet engine. Burning fuel expands gases, which turn a turbine, but the thrust is primarily provided by a propeller driven by the turbine.

**xenon** Gas that hardly ever reacts due to having a complete outer set of electrons. Used in powerful light bulbs.

# OXIDATION

Despite meaning "acid former," oxygen has no essential connection to acids. Oxidation is another of chemistry's misnomers; it needn't involve oxygen, although it may do. The key chemical fact in the combining of an element or compound with oxygen is that the substance relinquishes electrons to the oxygen atoms. Oxygen atoms are avid for electrons, but not uniquely so. Any substance that is stripped of some of its electrons by another substance that is more electron-hungry is said to be "oxidized." The recipient of the electrons, meanwhile, is "reduced." This combination of oxidation and reduction due to electron transfer is called a "redox" reaction, one of the most fundamental processes in chemistry. Combustion in oxygen is a special case of this process. Oxidation and reduction also occur at the two electrodes of a battery—electrons are transferred through the electrical circuit connecting them. Rusting of iron is an oxidation process in which the metal is converted to its reddish oxide. Energy changes in oxidation are not easy to generalize. One can imagine electrons "flowing downhill" to a lower-energy state—but the driving force of oxidation depends on other factors too. Oxidation of glucose by oxygen is the basis of respiration, but light energy can drive the reverse process of photosynthesis of sugars from oxygen and carbon dioxide.

**3-SECOND THRASH**
Oxidation is the process by which electrons are removed from a substance; in general, they are transferred to another substance in the complementary process of reduction.

**3-MINUTE THOUGHT**
Oxidizing agents, which are particularly good at oxidizing other substances by removing electrons, are essential components of explosives and rocket fuel. By enabling the oxidation reaction to happen rapidly, they can release chemical energy explosively. They often contain many oxygen atoms. Chlorate salts are like this, and sodium chlorate—a weed killer because of its ability to disrupt the redox processes of plant cells—makes an effective and potentially deadly homemade explosive when combined with sugar.

**RELATED TOPICS**
See also
CHEMICAL ENERGY
page 26

BATTERIES
page 74

BURNING
page 102

**3-SECOND BIOGRAPHIES**
JOSEPH PRIESTLEY
1733–1804
English chemist and one of the discoverers of oxygen, which he isolated as a gas and showed to be essential for respiration

NEIL BARTLETT
1932–2008
English chemist who explored oxidizing agents powerful enough to oxidize not just oxygen itself but also the inert gas xenon

**EXPERT**
Philip Ball

***Rust provides a very visual indication of oxidation as iron reacts with oxygen in the atmosphere to produce iron oxide.***

# BURNING

The controlled use of fire as an energy source seems to predate *Homo sapiens*—there is evidence that it was practiced by the earliest human species *Homo erectus* as much as 400,000 years ago, and perhaps even earlier. It's hard to be sure about the date, because distinguishing man-made from natural causes of burned stones and bones in the archeological record is challenging. It's sobering to realize that burning is still our main source of energy today: Oil, coal, and gas account for at least 85 percent of the world's energy resources. That's because they are relatively plentiful and inexpensive. However, these fossil fuels are only very slowly renewable—it typically takes millions of years for dead organisms to be converted geologically to coal and oil. What's more, burning the carbon-rich compounds produces carbon dioxide, the main greenhouse gas. Burning or combustion—the rapid combination of a substance with oxygen, releasing heat—is a complex chemical process, as Michael Faraday acknowledged in his historic 1848 lectures on "the chemical history of a candle." A simple candle flame has many chemical constituents, many of them highly reactive and ephemeral compounds called free radicals. Even today this burning process is not fully understood.

**3-SECOND THRASH**

Burning or combustion is the rapid combination of a substance with oxygen, releasing heat and usually light.

**3-MINUTE THOUGHT**

Simple combustion of carbon-rich substances can create complex molecules such as carbon cages, called fullerenes, and tubular structures called carbon nanotubes. These molecules are typically a nanometer (one-billionth of a meter) or so across, although nanotubes can reach a few centimeters in length. They are important in nanotechnology, the engineering of matter on nanometric scales. Fullerenes and nanotubes have surely existed on Earth as long as fire itself, and some have been seen in interstellar space.

**RELATED TOPICS**

See also
HEAT
page 18

CHEMICAL ENERGY
page 26

OXIDATION
page 100

**3-SECOND BIOGRAPHIES**

MICHAEL FARADAY
**1791–1867**
English scientist who made pioneering discoveries in electricity and used the burning of a candle to illustrate fundamental scientific principles to the public

GEORGE OLAH
**1927–2017**
Hungarian-American pioneer of hydrocarbon chemistry who studied ways of making hydrocarbon burning more efficient

**EXPERT**

Philip Ball

***Faraday used a candle flame to explore the energy reactions of fire, an oxidation reaction with potentially devastating results.***

# EXTERNAL COMBUSTION

**3-SECOND THRASH**
An external combustion engine takes energy from an external source of heat, such as the burning of coal, and converts it into mechanical work.

**3-MINUTE THOUGHT**
The efficiency of a heat engine can be quantified as the fraction of thermal energy converted to useful work. Practical steam engines tend to have low efficiency, often less than 10 percent. Early in the nineteenth century, Nicolas Carnot designed a theoretical engine with much higher efficiency, and showed that no other engine could exceed the efficiency of his design. This was not an idle boast, but a rigorous consequence of the second law of thermodynamics.

Heat engines are machines that convert thermal energy into useful work. Typically, they do this through the expansion and contraction of a working fluid as it is heated and cooled. In an external combustion engine, the source of heat is separate from the working fluid. The most familiar example is the steam engine, in which the working fluid is steam and the heat energy comes from burning coal. These engines operate on a repeating cycle, in which steam expands and contracts inside a cylinder to drive a piston. The earliest steam engines operated at low pressure and had to be extremely large to deliver a useful amount of power. While they were suitable for stationary applications, such as pumping, they could not be used for vehicle propulsion. However, by the end of the eighteenth century, smaller, more efficient high-pressure steam engines began to appear. Used in conjunction with a crankshaft to convert the back-and-forth motion of the piston into rotary motion, such engines soon revolutionized transportation. Steam traction dominated the railways throughout the nineteenth and well into the twentieth century, while huge ocean-going steamships ushered in an age of luxurious transatlantic travel.

**RELATED TOPICS**
See also
HEAT
page 18

INTERNAL COMBUSTION
page 106

THE SECOND LAW
page 142

**3-SECOND BIOGRAPHIES**
THOMAS NEWCOMEN
1664–1729
English engineer who invented the first practical steam engine

NICOLAS CARNOT
1796–1832
French physicist who developed the theory of heat engines

**EXPERT**
Andrew May

***Although heat engines had existed for some time, Carnot's work explained them scientifically and enabled more efficient engines to power the Industrial Revolution.***

# INTERNAL COMBUSTION

**3-SECOND THRASH**
Internal combustion engines, commonly used in cars and other forms of transport, produce mechanical power by burning a fuel mixture inside the engine.

**3-MINUTE THOUGHT**
One of the biggest concerns with internal combustion engines is the large quantity of pollutants they disgorge into the atmosphere. This includes both natural waste products from the combustion process, as well as particulate matter and gases resulting from incomplete combustion. Partly because they are so numerous, internal combustion engines are believed to be responsible for up to one-quarter of the world's anthropogenic carbon dioxide emissions and as much as one-third of smog-producing air pollution.

An internal combustion engine is a type of heat engine in which the working fluid is the fuel-air mixture itself, which is burned inside, rather than outside, the engine. While external combustion engines can use any readily available fuel, the fuel in an internal combustion engine is specific to the engine design—usually some form of refined oil. Combustion may occur continuously—as in a gas turbine, for example—or it may be intermittent, as in the four-stroke engines found in cars and other road vehicles. As with a steam engine, car engines involve a piston moving back and forth inside a cylinder. The repeating cycle in this case consists of four stages: intake, compression, combustion, and exhaust. Combustion may be initiated using an electrical spark, or—in a diesel engine—by producing sufficient heat during the compression stage to trigger spontaneous ignition. First appearing toward the end of the nineteenth century, and significantly smaller and more efficient than steam engines, internal combustion engines helped to kick-start the era of private motoring and heavier-than-air flight. By the end of the twentieth century, internal combustion—which encompasses rockets and jets as well as piston engines—was virtually ubiquitous in all forms of transport.

**RELATED TOPICS**
See also
OIL
page 62

EXTERNAL COMBUSTION
page 104

TURBINES
page 114

**3-SECOND BIOGRAPHIES**
NIKOLAUS OTTO
1832–91
German engineer who designed the first practical internal combustion engine

RUDOLF DIESEL
1858–1913
German engineer who invented the compression-ignition engine that bears his name

**EXPERT**
Andrew May

***Whether ignited by a spark in a gasoline engine or pressure in Diesel's variant, the internal combustion engine is a hallmark of the twentieth century.***

# ELECTROMAGNETISM

**3-SECOND THRASH**
Electromagnetism describes the interactions between electricity and magnetism, which can generate movement from electricity and electricity from movement.

**3-MINUTE THOUGHT**
Electromagnetic induction enables "step-down" transformers to reduce the high grid voltage into a lower-voltage domestic electricity supply. These transformers have two separate copper wire coils wound round a magnetic steel core. Alternating current (AC) from the grid flows through the "primary" coil producing an alternating magnetic field that induces AC in the "secondary" coil, which has fewer turns. Fewer turns in the secondary coil compared with the primary reduces the voltage, making it ready for use.

Much of the modern world is powered by electromagnetism, as it enables electric motors and electricity generators to work. The link between electricity and magnetism was reported in 1820 by Danish physicist Hans Christian Oersted. He saw a compass needle move when it was brought near a wire carrying an electric current, and so showed the current was creating a magnetic field around the wire. Today this effect is harnessed in electromagnets, which are only magnetic when current is flowing through their wire coils. A year later, English scientist Michael Faraday created the first electric motor. This consisted of a current-carrying wire that rotated around a fixed magnet thanks to the current in the wire creating its own magnetic field, which interacted with that of the stationary magnet. Modern electric motors work on the same principle, whether they are rotating an electric toothbrush or propelling an electric train. Electricity generators work in the opposite way. They convert movement into electricity via electromagnetic induction, in which a current is created in a wire when a magnet moves past it, or the wire moves through a magnetic field. This effect was independently discovered in 1830 by American physicist Joseph Henry and in 1831 by Michael Faraday, and enables power stations and wind farms to generate electricity.

**RELATED TOPICS**
See also
DOWN THE WIRE
page 92

TURBINES
page 114

WIND
page 124

**3-SECOND BIOGRAPHIES**

ANDRÉ-MARIE AMPÈRE
1775–1836
French physicist and mathematician who developed a mathematical theory that explained the basic effects of electromagnetism

JOSEPH HENRY & MICHAEL FARADAY
1797–1878 & 1791–1867
American physicist and English physicist and chemist who independently discovered electromagnetic induction in 1830 and 1831 respectively

**EXPERT**
Sharon Ann Holgate

***Ampère's theoretical work encouraged Faraday to investigate electromagnetism, creating motors and generators.***

N
S
−
+

# QED

**3-SECOND THRASH**
Light energy in photons is absorbed by electrons in atoms and re-emitted; this is central to the interactions of light and matter, and of matter with other matter.

**3-MINUTE THOUGHT**
QED has its share of complex mathematical equations, but it is unusual in physics in also having a mechanism of representation that is intuitive and simple. This involves Feynman diagrams, named after their charismatic inventor, Richard Feynman. The diagrams represent the different ways that particles can interact—photons are shown as wiggly lines and matter particles as straight lines. Although apparently simplistic, the diagrams are used as the basis for structuring calculations and approximations.

An essential conversion of energy in nature involves the interaction of light and matter, often given the acronym QED for "quantum electrodynamics." QED operates at the level of quantum particles, something that was originally demonstrated in the photoelectric effect. In a typical QED interaction, an incoming photon of light is absorbed by one of the electrons that is part of an atom. The energy that was in the photon is converted into potential energy in the electron, which now occupies a higher orbit around the nucleus. This mechanism became the foundation of quantum physics, as the electron cannot occupy orbits of all possible potential energies, but only specific fixed levels, between which it jumps in a quantum leap. Often the electron will drop back down to a lower level at a later time, giving off a photon of light energy in the process. Even when we don't see light involved, such exchanges are constantly happening. For example, when two pieces of matter interact—when you sit on a chair, for instance—photons are exchanged between the electrons in the atoms of the two objects, carrying the electromagnetic force that means that you sit on a chair rather than sliding straight through it.

**RELATED TOPICS**
See also
POTENTIAL ENERGY
page 22

RADIATION
page 84

SOLAR
page 122

**3-SECOND BIOGRAPHIES**
PAUL DIRAC
1902–84
English physicist who laid the groundwork for QED

RICHARD FEYNMAN
1918–88
American physicist who won the 1965 Nobel Prize for physics for his work on QED alongside Julian Schwinger and Sin'ichirō Tomonaga

**EXPERT**
Brian Clegg

***Photons of light act as the carriers of electromagnetic energy, for example between the atoms in a person and the atoms in a chair, allowing the chair to provide support.***

1854
Born in London, UK

1877
Graduates from Cambridge University

1883
Marries Katharine Bethell

1884
Develops turbine engine at Clarke, Chapman and Co.

1889
Founds C.A. Parsons and Co.

1890
Parsons' Newcastle and District Lighting Company builds world's first turbine power station

1897
Demonstrates first turbine-powered vessel, *Turbinia*, faster than any Royal Navy ship

1898
Elected fellow of the Royal Society

1911
Knighted by George V

1925
Follows his father's interest in astronomy by buying the Grubb Telescope Company, which became Grubb Parsons

1929
Awarded the Order of Merit

1931
Dies in Kingston, Jamaica

# CHARLES ALGERNON PARSONS

When we consider the origins of electricity generation, it is natural to think of Michael Faraday and the device that converts mechanical energy into electrical. However, for more than 100 years, the standard mechanism for producing such mechanical energy from the chemical energy of fuel via heat has been the steam turbine, invented in its modern form by Charles Algernon Parsons.

In the history of science, Parsons tends to be overshadowed by his father William, who built the then largest telescope in the world, the "Leviathan of Parsonstown," at his home Birr Castle in Ireland. However, Charles has had a far larger impact on our everyday lives.

The youngest of four sons, Charles Parsons studied at Trinity College, Dublin, and then St. John's College, Cambridge. At the time, the youngest son of an earl was likely to go into the clergy or the military, but Parsons, fascinated by mechanical and electrical engineering, took an apprenticeship at W. G. Armstrong (later Armstrong Whitworth) in Newcastle. After some time working in military and electrical engineering, while at Clarke, Chapman and Co., Armstrong developed an effective steam turbine, using multiple sets of turbine blades to break down the steam pressure to manageable levels. Although Parsons' invention was predated by a turbine produced by the Swedish engineer Gustaf de Laval, the way that Parsons designed his device with multiple blades made all the difference for practical usability.

Soon after, Parsons set up his own firm, C. A. Parsons and Co., based in Newcastle, to produce turbine-driven generators. He would go on to produce turbine engines for ships, with the first liners to use them going into service in 1905 and the first battleship in 1906. However, this use is dwarfed in importance by the role that steam turbines play in electricity generation. Whether the power station is using coal, gas, oil, or nuclear power, the mechanism for turning the heat produced by the furnace or reactor into electrical generation remains the steam turbine. And Charles Algernon Parsons is the man behind the technology. His Heaton factory was taken over by Siemens Energy, and still runs as the C. A. Parsons Works.

Although Parsons' place of death is given as Kingston, Jamaica, he died while on a Caribbean cruise with his wife. His body was brought back to London, with a memorial service held in Westminster Abbey, before being buried at his local parish church in the village of Kirkwhelpington, Northumberland.

*Brian Clegg*

# TURBINES

A turbine is a device that employs rotary motion to convert energy from one form or another into usable work. Early precursors include windmills and waterwheels, and in their modern forms both wind and water turbines are used to generate electricity. Also widespread in this context are steam turbines, which obtain their energy from heat—either through the burning of coal or oil, or from a nuclear reactor. As with any steam engine, a steam turbine involves external combustion. In contrast, a gas turbine relies on internal combustion of a fuel-air mixture. In engines of this type, the shaft of the turbine is used to drive a large fanlike compressor that draws air into the combustion chamber. The shaft can also be used to perform external work—for example, by driving the propeller of a ship or a "turboprop" aircraft. However, if there is no external load on the shaft, most of the energy generated by the engine goes into pushing out the exhaust plume—the hot gases from the turbine mixed with cooler airflow from the compressor. This results in a net thrust, and is the principle behind the "turbofan" jet engines found on most modern airliners and military aircraft.

**3-SECOND THRASH**

Turbines produce power via rotary motion; wind, water, and steam turbines are used in electricity generation, while gas turbines are found in ships and aircraft.

**3-MINUTE THOUGHT**

Most modern diesel cars are "turbodiesels," which employ a small turbine—called a turbocharger—designed to boost the engine's power and efficiency by increasing the amount of air inside the cylinders prior to combustion. In effect, a turbocharger is a miniature jet engine that uses some of the waste energy from the car's exhaust gases to drive a rotating compressor fan. This fan then forces air into the engine at higher than atmospheric pressure.

**RELATED TOPICS**

See also
EXTERNAL COMBUSTION
page 104

INTERNAL COMBUSTION
page 106

WIND
page 124

**3-SECOND BIOGRAPHIES**

CHARLES ALGERNON PARSONS
1854–1931
English engineer who designed and built the first steam turbine

FRANK WHITTLE
1907–96
English engineer who invented the jet engine, a gas turbine suited to aircraft propulsion

**EXPERT**

Andrew May

***The turbine was first designed to employ steam to produce power, but would also prove an essential component in the design of the jet engine.***

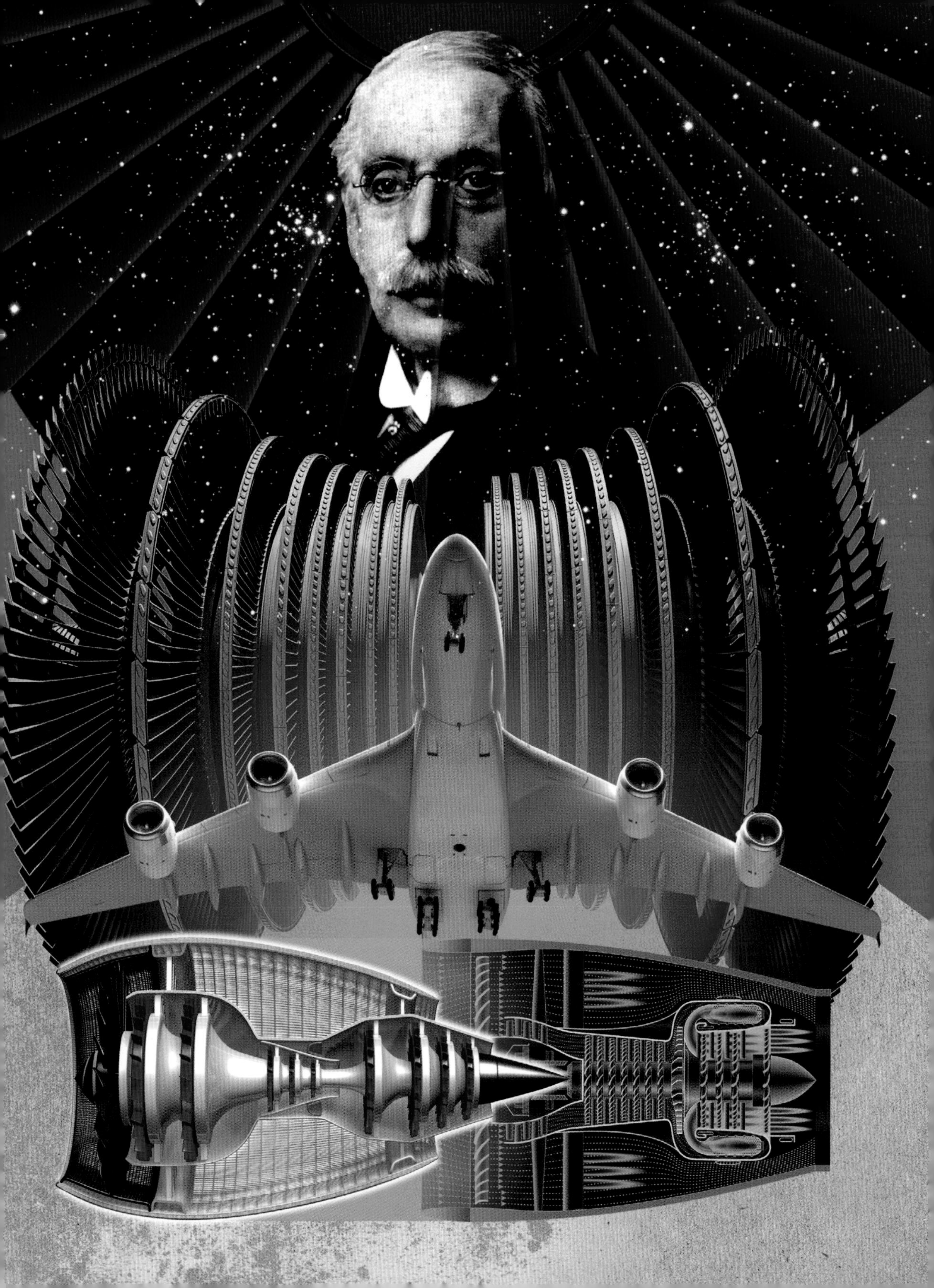

# GOING GREEN

**absolute temperature** Temperature scales such as Celsius are based on arbitrary fixed points such as the freezing and boiling points of water. An absolute scale starts at the lowest possible temperature, absolute zero.

**chemical feedstocks** Source materials for the chemical constituents of a product, for example the hydrocarbons used to produce plastics.

**Chernobyl accident** In April 1986, an explosion at the Chernobyl nuclear power station in Ukraine resulted in significant radioactive fallout across Europe.

**dry steam field** Steam is usually a mixture of water vapor and hot liquid droplets. However, superheated steam, which is above water's boiling point, has no water droplets and is sometimes called dry steam. Natural geysers producing superheated steam form a dry steam field.

**Earth crust, mantle and core** The Earth has three major layers: The crust, 3–47 miles (5–75 km) thick; the mantle, around 1,740 miles (2,800 km) deep; and the core, extending around 2,175 miles (3,500 km) from the Earth's center. The core is mostly molten metal with a solid inner section.

**energy density** Amount of usable energy per unit mass of an energy storage product.

**Fukushima** In March 2011, a tsunami damaged the coolant system of a nuclear reactor in Fukushima, Japan, causing meltdowns and release of radioactivity.

**greenhouse and toxic gases underground** When water hits molten rock underground it produces steam that can be used to drive turbines. Such steam may contain greenhouse gases such as methane and nitrous oxide, and toxic gases such as hydrogen sulfide and carbon monoxide.

**heat pump** Device that uses energy to transfer heat from a cooler to a hotter place, for example a refrigerator.

**magma chamber** Large volume of underground molten rock. Under high pressure, it can force itself upward, causing a volcanic eruption.

**nuclear chain reaction** In nuclear fission, the nucleus of an atom is hit by a neutron, splitting into two. If this reaction produces two or more neutrons, each can split another nucleus, producing a self-sustaining chain of reactions.

**nuclear fission** Splitting of the central nucleus of a heavy atom, such as uranium, producing lighter elements, neutrons, and energy. This is the energy source of all current nuclear power plants.

**nuclear fusion** Merging of two or more light atomic nuclei, such as hydrogen, to produce a heavier nucleus and energy. This is the energy source of the Sun and experimental fusion reactors.

**nuclear power and nuclear weapons** All early nuclear power stations were designed to produce material for nuclear weapons. This has meant that development has focused on reactors that are not the best for generation purposes.

**pebble-bed reactor** Alternative design for nuclear reactors in which the fuel is provided in tennis-ball sized spherical "pebbles." Should the system overheat, it stops producing the slow neutrons needed for a nuclear reaction, automatically shutting down.

**quantum theory** Physics theory describing the behavior of the very small particles that make up matter and light.

**reaction nozzle** Simple turbine, in which the rotor is turned by squirting out liquid at an angle to the direction of motion. It is most familiar in lawn sprinklers.

**Salter's duck** Officially the Edinburgh duck, an early device for converting the energy of waves into electricity.

**second law of thermodynamics** Sometimes paraphrased as "things run down," the second law says that in an isolated system, when two bodies are in contact, heat flows from the hotter to the colder. This can be phrased as entropy—the level of disorder in the system—stays the same or increases.

**tectonic plates** Large-scale plates of rock covering the surface of the Earth that gradually move, causing mountains to form at their boundaries, also triggering earthquakes and volcanoes.

**tidal energy** Energy production using the motion of the water due to tides.

# BIOFUELS

**3-SECOND THRASH**
Biofuel is any organic matter turned into a useful fuel that can be grown quickly—that is, within the space of about a year.

**3-MINUTE THOUGHT**
Biofuels have several advantages over other fuels. They can be grown around the world and suit different levels of technological development, they're almost carbon neutral, and they can be used as chemical feedstocks for materials such as plastics and pharmaceuticals. But there are drawbacks: Biofuels absorb nutrients from the soil, which means a greater need for fertilizers, and they require a lot of water. Also, they use land that might instead have been useful for growing food crops.

Biomass is any organic material that can be considered a source of energy. This may be because it can be immediately burned (for example, wood) or converted into useful fuels (for example, crops grown for liquid biofuels such as ethanol and biodiesel). While fossil fuels can be considered fossilized biomass, in common usage the term refers to matter that can be produced relatively quickly (rarely taking more than a year) and so is referred to as a renewable resource. Biomass is the primary source of energy for over half the planet's population even though it provides only about one-eighth of the world's energy needs. The most significant biofuel is ethanol, which can be made by fermenting the sugars in any plant—corn is typically used. Today, ethanol is often added to gasoline; this blend is identified by the letter "E" with a number signifying the percentage after it. For example, E10 gasoline contains 10 percent ethanol. Unfortunately, ethanol has roughly half the energy density of gasoline. Similarly, biodiesel, typically formed from oils extracted from crops such as soybeans, sunflower seeds, or rapeseed, can be blended with diesel in small amounts (B5 = 5 percent biodiesel). Crops aren't the only possible source of biofuels. Research is currently taking place into sourcing them from fungi and gut bacteria.

**RELATED TOPICS**
See also
CHEMICAL ENERGY
page 26

COAL
page 60

OIL
page 62

**3-SECOND BIOGRAPHY**
HENRY FORD
1863–1947
American inventor and industrialist who declared ethanol to be the "fuel of the future;" his Model T was designed to run on ethanol and gasoline

**EXPERT**
Simon Flynn

***Although we traditionally produce carbon fuels from fossil sources, biofuels use crops to take carbon from the atmosphere.***

# SOLAR

**3-SECOND THRASH**
Solar energy is readily available in large quantities during daylight, but at the moment there is a trade-off between efficiency and cost for photovoltaic solar cells.

**3-MINUTE THOUGHT**
Some have suggested that the best long-term solar solution is to assemble vast solar panels in space, where there is no weather to reduce the efficiency of conversion. The problem then is how to get the energy back down to Earth. The proposal is to do this by firing intense beams of microwaves through the atmosphere—which would then be converted back to electricity at a ground station. This inevitably has significant practical and safety issues.

Around 90 billion megawatts of solar power hits the Earth, more than 7,000 times total global consumption. All we need to do is get some of that energy into a usable form. This often means using photovoltaic solar cells, converting sunlight to electricity—but we can use the light more directly, from the domestic heating approach of putting tubes of water into sunlight, to high-tech solar plants, where arrays of mirrors focus the Sun's energy to heat water, or nitrate salts that can reach 1,112°F (600°C). However, photovoltaic cells remain the core approach to harnessing solar energy. These rely on the light boosting the energy of electrons in special materials, which release those electrons as a flow of electricity. As the costs drop and efficiency of converting light energy into electricity rises, solar cells become more practical. It has been suggested that the whole of Europe could be supplied by solar farms in North Africa, using high-voltage direct current (DC) transmission to carry the energy to consumer countries, but such a supply would be politically risky. Most countries prefer a mix of sources, and the limit of sunlight to daytime means solar could only be dominant with advanced storage technologies.

**RELATED TOPICS**
See also
BATTERIES
page 74

RADIATION
page 84

QED
page 110

**3-SECOND BIOGRAPHIES**
WILLIAM HERSCHEL
**1738–1822**
German-born British astronomer who discovered infrared radiation in sunlight

CHARLES FRITTS
**1850–1903**
American inventor who devised the first working selenium photovoltaic cell in 1883

ALBERT EINSTEIN
**1879–1955**
German physicist whose paper explaining the photoelectric effect won him the Nobel Prize

**EXPERT**
Brian Clegg

***The Sun provides more energy than we are ever likely to need, with a range of technologies able to capure it.***

# WIND

Wind generation is one of the most visible sources of green energy, but it engenders controversy. The theory is good—there is "free" energy whenever the wind is blowing. An individual turbine can generate 3–5 megawatts (MW), with large windfarms producing around 500–1,000 MW, comparable in output to a large conventional power station. However, there is one practical issue and three environmental concerns. Practically, unless there are large-scale storage facilities to hold energy until needed, wind power's inconsistency means it will always need back-up sources. Of the environmental concerns, the danger to birdlife is perhaps exaggerated. Although wind turbines do sustain bird strikes, a turbine causes fewer bird injuries than a single cat. More significant are visual and noise pollution. These 200–300 feet (60–90 m) towers with twin blades are hard to ignore and many think they spoil natural landscapes. Although this problem can be overcome by siting turbines offshore, this raises the cost significantly. Noise pollution is also a genuine problem for those living close to a turbine, meaning that siting needs to be carefully judged. Although wind generation has dropped somewhat in favor, its contribution to world energy doubles every three years and it remains an important contributor to a sustainable green energy balance.

**3-SECOND THRASH**
A wind farm can produce comparable energy to a conventional power station and is low-carbon, but output is more variable and the technology has environmental issues.

**3-MINUTE THOUGHT**
Wind power is, in effect, redirected solar energy, with the weather system acting as an intermediary. Energy from sunlight differentially heats different parts of the atmosphere. Air molecules are always zooming around, but wind results from concerted movement. As warmer air rises and cooler air sinks, masses of air begins to move. Combine this with the effect of the planet's rotation and the result is that air begins to travel from place to place as wind.

**RELATED TOPICS**

See also
KINETIC ENERGY
page 20

TURBINES
page 114

SOLAR
page 122


**3-SECOND BIOGRAPHIES**
HERO OF ALEXANDRIA
**fl. c. 62 CE**
Greek engineer who used a wind turbine to power an organ

JAMES BLYTH
**1839–1906**
Scottish engineer who produced the first electricity-generating wind turbine

POUL LA COUR
**1846–1908**
Danish inventor who constructed the first wind generator to be used for domestic power

**EXPERT**
Brian Clegg

***These present-day windmills capture otherwise unused natural energy and can be sited offshore to minimise visual impact.***

# HYDRO

Of all the green sources of energy, hydropower has been used for the longest time, predating even wind power. Initially hydropower was a purely mechanical concept. The kinetic energy of streams and rivers, derived from potential energy as the water ran downhill, was turned into mechanical work to power mills and saws. Hydropower then took the lead in electricity generation; the first hydroelectric power station at Lord Armstrong's house, Cragside in Northumberland, began generating electricity in 1878. The dramatic force of Niagara Falls was brought online as an electricity generator just one year later. For a considerable period of the twentieth century, hydroelectric power stations, making use of dams to produce their potential energy, were the only significant renewable electricity generation source. These projects have become less common in the West in recent years, as the environmental and social impact of projects that can involve leaving whole towns underwater have made giant dams controversial. However, development has continued, for example, in China, where the Three Gorges Dam in Hubei province opened between 2003 and 2012, gradually adding capacity to reach an output of 22,500 megawatts, comparable to around 30 conventional power stations.

**3-SECOND THRASH**
Hydropower uses the kinetic energy of moving water to push wheels, creating usable energy. Some of the largest single power stations are now hydro-based.

**3-MINUTE THOUGHT**
For hydropower you need a body of water higher than its surroundings, so that the potential energy of the water—which is converted to kinetic energy as it runs downward under the pull of gravity—can power a turbine. The energy source that makes this possible is the Sun: it evaporates water from sea level and powers its transfer by convection through the atmosphere, followed by precipitation as rain at higher altitudes.

**RELATED TOPICS**
See also
POTENTIAL ENERGY
page 22

GRAVITY
page 38

WATER STORAGE
page 70

**3-SECOND BIOGRAPHIES**
JOHANN SEGNER
1704–77
Hungarian scientist who devised the first practical water turbine based on reaction nozzles

JACOB SCHOELLKOPF
1819–99
German-American industrialist behind the Niagara Falls hydroelectric power station

**EXPERT**
Brian Clegg

***Segner's eighteenth-century water turbine was a simple device, but large-scale hydroelectric power plants can produce the same output as multiple conventional power stations.***

**1824**
Born in Belfast, UK; his father, James Thomson, subsequently becomes professor of mathematics at Glasgow University

**1845**
Graduates in mathematics from Cambridge University

**1846**
Appointed to the chair of natural philosophy at Glasgow University

**1857**
Sails on an unsuccessful expedition to lay the first transatlantic telegraph cable

**1864**
Estimates the age of Earth at 20 million to 400 million years, which seems too short for Darwinian evolution; he later refined his estimation to between 20 and 40 million years

**1866**
Knighted by Queen Victoria for his services to the transatlantic cable project, which had now succeeded

**1867**
Publishes, with Peter Guthrie Tait, the highly influential textbook *Treatise on Natural Philosophy*

**1892**
Becomes Baron Kelvin of Largs in Ayrshire

**1907**
Dies in Largs, aged 83

# WILLIAM THOMSON, LORD KELVIN

William Thomson was many things: a mathematician, physicist (he did much to establish physics as a discipline), and engineer, but also a public celebrity and the first British scientist to be made a lord. Born in Belfast of Scottish heritage, he studied mathematics at Cambridge before becoming professor of natural philosophy in Glasgow University.

Thomson became famous partly for his expertise in electricity and electrical technology, advising on the laying of the first transatlantic telegraph cables in the 1850s and 1860s. That operation was dogged by technical problems, but its eventual success led to Thomson's knighthood and to public adulation.

But it was in the new science of thermodynamics that Thomson made his greatest contributions to science. He was one of the first to appreciate the enormous, even cosmic, ramifications of the discipline. He proposed an absolute temperature scale (the units of which are now named after him: Kelvin), interpreted heat as a form of motion, and laid the foundations of the second law of thermodynamics.

All that did not obscure Thomson's appreciation of the practical side of thermodynamics. It was, after all, devised to help explain the engines of the Industrial Revolution, and in 1852 Thomson himself proposed a new kind of energy-generating device: the heat pump. This transfers ("pumps") heat energy against its normal direction, absorbing it from a cooler source and moving it to a warmer sink. This process, familiar from refrigerators and air conditioners, requires an external source of power. The device can be run in reverse to produce heating, and Thomson, who described the thermodynamic theory of such a device, hoped that it might be used both to warm and to cool houses.

Thomson illustrates the hazards of becoming a "public scientist." In his later years he was fêted and consulted by the press as an expert, while looking increasingly out of touch to his peers. He is often remembered now as someone always just behind the curve of *fin-de-siècle* science. His huge underestimation of the age of the Earth, based on a calculation of its rate of heat loss, came just as the discovery of radioactivity supplied a hitherto unsuspected heat source. He doubted that aviation would be possible. Although the popular notion that he pronounced physics finished except for increasing precision—just as quantum theory was about to explode on the scene—seems to be false, it fits with the image of the older Thomson as a man no longer in touch with the discipline he helped to launch.

***Philip Ball***

# WAVES

**3-SECOND THRASH**
Wave power provides a green source of energy that is weather-dependent but less variable than wind or solar; at the moment, outputs are relatively small.

**3-MINUTE THOUGHT**
Another significant seawater energy source is the tide—along with geothermal and nuclear, one of the few energy sources that isn't directly or indirectly powered by sunlight, but rather by the gravitational pull of the Moon and Sun. Tidal stations are often built or planned across bays and estuaries, typically providing higher output than waves—the proposed UK Severn barrage would generate an average output of around 2,000 MW—but tend to have higher environmental impact than wave stations.

Seawater is constantly in motion due to the wind passing over its surface—and, like the movement of water harnessed by hydropower, this can be used to generate electricity. Wave power is one of the most recent green technologies to become practical. It wasn't until the end of the twentieth century that wave energy became a potentially cost-effective source. Because the technology has yet to reach the mass-production status of, say, wind turbines, it remains expensive, and there are still many experimental technologies in play. These divide into systems that rely on the up-and-down movement of a floating device to harness kinetic energy; systems in which waves impact a device causing side-to-side motion; and systems in which the wave is used to transfer water into a reservoir, producing potential energy to be harnessed. Although the energy in waves varies considerably with the weather, it is rare that there is no wave energy available, making this approach more consistent than some green sources. But as yet the output is limited. The world's largest wave farm off Scotland started at 3 MW output and will eventually reach 40 MW; compare this with the largest solar farms at around 600 MW and wind farms of up to 6,000 MW.

**RELATED TOPICS**
See also
KINETIC ENERGY
page 20

TURBINES
page 114

HYDRO
page 126

**3-SECOND BIOGRAPHIES**
YOSHIO MASUDA
1925–2009
Japanese naval expert who was the first major proponent of wave energy

STEPHEN SALTER
born 1938
South African-born engineer whose "Salter's duck" was one of the first practical designs for wave energy collection

**EXPERT**
Brian Clegg

***Tidal and wave-powered generators can be costly but they are less intermittent renewable energy resources than wind or air.***

SMITH ISLAND SHOAL
CHESAPEAKE LIGHTSHIP
Match Line

# GEOTHERMAL

Geothermal energy is provided by heat stored in the Earth. At shallow depths, say 10–15 feet (3–4.5 m), a temperature of 50–59°F (10–15°C) is maintained. Heat pumps enable this constant temperature to be used to cool buildings in the summer and warm them during the winter by transferring heat from hot to cold. Any miner will tell you that as you dig deeper the temperature increases. In the upper part of the Earth's crust, and away from tectonic plate boundaries, temperature increases on average by 45–54°F (25–30°C) per kilometer depth. Higher temperature gradients occur where tectonic plates meet (for example, in Iceland) or in places that sit on top of large magma chambers (such as Yellowstone Park in the United States). In locations like these, steam may be used to produce electricity by driving turbines. This can be achieved because of the existence of underground reservoirs of steam or water above 212°F (100°C), which is kept in liquid form because of extreme pressure. As the latter gets nearer the Earth's surface, pressure drops and it turns into steam. These reservoirs can often be replenished, making geothermal a renewable resource. Issues include the possibility of an increase in the release of greenhouse and toxic gases stored underground and the fact that the power plants are location-specific.

**3-SECOND THRASH**
Geothermal energy, heat stored within Earth, can be used to warm and cool buildings anywhere and to produce electricity.

**3-MINUTE THOUGHT**
The temperature of the Earth's core is about 10,832°F (6,000°C), similar to the Sun's surface. As heat always transfers from hot to cold, this results in energy flowing outward to the Earth's surface. A legacy of the Earth's formation, this temperature was very probably even higher 4 billion years ago. Another source of energy is the radioactive decay of potassium-40, thorium-232, uranium-235, and uranium-238, found in the Earth's crust, mantle, and core.

**RELATED TOPICS**
See also
HEAT
page 18

NUCLEAR ENERGY
page 28

FISSION
page 64

**3-SECOND BIOGRAPHY**
PIERO GINORI CONTI
1865–1939
Italian businessman who, in 1904, built the first geothermal power generator, which used a dry steam field in Tuscany

**EXPERT**
Simon Flynn

***Earth's interior gets hotter toward the core, enabling pipes sunk deep into the ground to tap into a natural source of heat.***

# NUCLEAR

**3-SECOND THRASH**
Nuclear power is green because, unlike fossil fuels, it does not generate greenhouse gases; but fission is increasingly politically sensitive, while practical fusion is decades away.

**3-MINUTE THOUGHT**
An alternative to the traditional uranium-235 fission plant is a reactor based on thorium. Mined thorium-232 is treated (potentially by the same reactor) to produce uranium-233, which splits when hit by a neuron, producing further neurons. Thorium is more abundant than uranium, produces far less radioactive waste in the process, and can be made meltdown-proof. But development has been slow because such reactors aren't good sources of materials for nuclear weapons, meaning that funding has been limited.

Nuclear power is the most controversial energy source considered green. It is environmentally friendly in the sense that its use reduces greenhouse gas emissions, but the need to store waste over extremely long timescales, and the fact that it tends to create fear—related to the association of nuclear power stations with nuclear weapons, and reactor accidents at Chernobyl and Fukushima—mean that many green organizations campaign *against* nuclear power. However, some environmental experts, such as James Lovelock, argue forcibly that nuclear power (alongside resources such as wind and waves) is the only option that will enable us to move away from coal and gas. Many of the problems with existing fusion reactors reflect their aging design. For example, there is a form of nuclear plant known as a pebble-bed reactor that is inherently incapable of meltdown. But developing new nuclear power stations is extremely expensive and at the time of writing faces considerable resistance. Current nuclear reactors make use of nuclear fission, but the greenest option is, without doubt, nuclear fusion, the power source of the Sun, which produces far less waste. It is highly unlikely, however, that we will see a commercial nuclear fusion reactor before the middle of the twenty-first century.

**RELATED TOPICS**
See also
NUCLEAR ENERGY
page 28

FISSION
page 64

FUSION
page 66

**3-SECOND BIOGRAPHIES**

GEORGE PAGET THOMSON
1892–1975
English physicist who, with Moses Blackman, had the first patent on a nuclear fusion reactor

LEO SZILARD
1898–1964
Hungarian physicist who realized the practicality of the nuclear chain reaction

**EXPERT**
Brian Clegg

***Traditional nuclear reactors are green on carbon emissions but produce dangerous waste; in the future, fusion reactors offer the potential of cleaner, safer power plants.***

# ENERGY & ENTROPY

## ENERGY & ENTROPY
GLOSSARY

**applied thermodynamics** Thermodynamics is the study of the movement of heat. Its practical application is usually to devices that use heat to produce work, such as steam turbines.

**big bang** Best accepted theory of the origin of the universe. The big bang is strictly the point in time at which the universe began to expand.

**black hole** A star that has undergone gravitational collapse to become a dimensionless point. Although not detectable directly, many apparent black holes have been discovered from their effect on their surroundings.

**Boltzmann's constant** A key value in understanding the behavior of gases, the Boltzmann constant, which is around $1.38064853 \times 10^{-23}$ J/K (joules per kilogram), fixes the relationship between the pressure, volume, and temperature of a gas and the number of molecules present.

**bomb calorimeter** A calorimeter measures the amount of heat produced by a reaction. In a bomb calorimeter, fuel is burned in an enclosed container (usually pressurized), heating water to measure the energy released.

**closed system** System with no matter flowing into it or out from it—although heat can be exchanged with other parts of the universe. Confusingly, the term is also used for an isolated system.

**cosmic inflation** A mechanism to explain why the universe appears to be more uniform than the big bang theory predicted, inflation was a vast increase in the volume of the universe that took place in a fraction of a second, around $10^{-35}$ seconds into the existence of the universe.

**dark energy** The expansion of the universe is accelerating. Something must be powering this acceleration—this unknown "something" is called dark energy. The total dark energy appears to be around 68 percent of all mass/energy.

**entropy** Measure of the disorder in a system, important in understanding heat, its distribution, and its use. Entropy can be statistically determined as the number of distinguishable different ways to arrange the component parts of a system.

**equilibrium state** When a system is in thermodynamic equilibrium, internal heat flows balance out, so there is no net flow from one place to another. The existence of stars and other generators of heat mean that the universe is not in equilibrium.

**isolated system** System in which no energy or matter flows in or out—it is entirely isolated from the universe around it.

**lower energy quantum state** Quantum systems, such as atoms, give off energy when they drop to a lower quantum state. It is theoretically possible (although unlikely) that the apparent minimum energy level of the universe is a plateau, stable but not the actual lowest state. If so, quantum effects could cause the universe to drop catastrophically to an even lower energy state.

**randomness/disorder** Randomness is a lack of detectable patterns. In entropy, disorder reflects the number of ways an object's component parts can be organized—the more different ways, the higher the disorder. For example, the letters in a specific book are organized in a particular way, but if the letters were scrambled up there would be very many more ways to arrange them, making them more disordered.

**Rankine cycle/engine** Mechanism of steam turbines, such as those in most power stations, used to convert heat into work. It was devised by Scottish engineer William Rankine.

**spacetime** German physicist Hermann Minkowski devised the concept of spacetime, a four-dimensional combination of space and time that is essential to understanding Einstein's special theory of relativity.

**statistical mechanics** Mostly used to predict the thermodynamic behavior of gases, statistical mechanics takes a probabilistic view of the behavior of a system averaged across its many different components.

**time's arrow** The "direction" of time seems to point toward the future. This is thought to be the result of thermodynamics and the requirement for entropy to stay the same or increase.

# WHAT IS ENTROPY?

**3-SECOND THRASH**
Entropy measures the number of different states a system can occupy and is the reason why some energy in machines and engines is always lost.

**3-MINUTE THOUGHT**
Even black holes have entropy. Black holes are regions of spacetime with gravitational fields so strong that nothing, including light, can escape. Physicists Stephen Hawking and Jacob Bekenstein showed that entropy is proportional to the surface area of a black hole's event horizon (or boundary). So as a black hole gobbles up more stuff, it becomes more massive and expands, growing its event horizon surface area and so increasing its entropy.

Entropy is often described as a measure of the disorder in a system—the more entropy, the higher the disorder. It is the reason why not all energy can be converted into useful work and why some energy is always "lost" in machines—why, for example, a car wastes energy on heating the engine, generating noise, and vibrating the bodywork rather than using it all to propel the vehicle forward. Entropy is a statistical concept, describing a system as a whole rather than its individual components. We can measure the volume, temperature, pressure, and energy of a system and calculate its entropy even though we do not know exactly how all the atoms and molecules within are individually moving. Unlike energy, entropy can be created. Entropy is a measure of all the different possible ways that a system can be arranged at a microscopic level. As the number of particles within a system increases, or the space available for them to move about in expands, or the energy within the system grows, the number of possible different states of the system increases, and so entropy increases. Machines work by constraining the number of possible states—or entropy—within them, which means that some entropy must escape, taking with it energy as, say, noise, vibration, and waste heat.

**RELATED TOPICS**
See also
THE SECOND LAW
page 142

INCREASING DISORDER
page 144

THE CLOSED SYSTEM
page 146

**3-SECOND BIOGRAPHIES**
RUDOLF CLAUSIUS
1822–88
German physicist who first described this unavoidable energy dissipation mathematically and gave it the name entropy

LUDWIG BOLTZMANN
1844–1906
Austrian physicist who underpinned the concept of entropy in statistical terms and showed that disorder tends toward a maximum

**EXPERT**
Leon Clifford

***Entropy can be reduced, but always at the cost of energy—in a machine where entropy is constrained, energy will always be lost in the process.***

# THE SECOND LAW

**3-SECOND THRASH**
The second law of thermodynamics explains why your coffee gets cold and why many natural processes tend to only run in one direction.

**3-MINUTE THOUGHT**
Does the second law of thermodynamics apply to our universe? If so, then the universe may end in "heat death." Matter will decompose into its constituent particles and energy will be dissipated among those particles, which will move randomly through space for eternity. Everything will cool to the same temperature—a minute fraction of a degree above absolute zero. And it will stay that way. In utter darkness. Forever.

In thermodynamics, the second law explains why heat flows from hot objects to their cooler surroundings and not vice versa. It is the reason why the contents of a refrigerator would warm to room temperature if its motor stopped pumping out heat to maintain a temperature difference between the inside of the fridge and the air around. In scientific terms, the second law states that the total entropy of an isolated system—the number of possible states that a system may occupy—stays the same or increases over time. It means that entropy can be created even though energy cannot. It also means that as the entropy of a system increases so the energy within that system tends to get dissipated, spread out, and shared between all the components of that system rather than remaining concentrated in particular places. This is why hot drinks surrender heat and grow cool while cold drinks absorb heat and grow warm, until they reach the same temperature as the surrounding air. The second law also implies that natural processes tend to go in only one direction and, left to themselves, are irreversible. So hot objects in cooler surroundings invariably cool rather than warm, balloons naturally deflate rather than spontaneously inflate, and fire inevitably converts wood into flames, ash, and smoke and not the other way round.

**RELATED TOPICS**
See also
WHAT IS ENTROPY?
page 140

INCREASING DISORDER
page 144

THE CLOSED SYSTEM
page 146

**3-SECOND BIOGRAPHIES**
RUDOLF CLAUSIUS
& LORD KELVIN
**1822–88 & 1824–1907**
German and Scottish-Irish physicists who independently formulated equivalent definitions of the second law of thermodynamics

JAMES CLERK MAXWELL
**1831–79**
Scottish physicist who conceived a thought experiment suggesting that the second law could be violated

**EXPERT**
Leon Clifford

***At its simplest, the second law of thermodynamics tells us that heat flows from hotter to cooler bodies.***

# INCREASING DISORDER

**3-SECOND THRASH**
Entropy is equivalent to randomness or disorder, so growing entropy means increasing disorder.

**3-MINUTE THOUGHT**
The seemingly inevitable drift toward increasing randomness and disorder in the universe may explain why we experience time as running forward and not backward—the so-called arrow of time. However, this apparent one-way flow may be a feature of our existence at the macroscopic level and may not hold true at very small scales, where thermodynamics meets quantum mechanics. In the quantum world, the direction of time is less obvious.

One way of thinking about entropy is in terms of order and disorder, or randomness. So, for example, an ice cube is a highly ordered lattice of water molecules but the puddle left after it has melted is a pool of randomly moving water molecules and is much less ordered. As entropy increases, order decreases and the amount of disorder, or randomness, grows. Since the second law of thermodynamics says that entropy increases as energy is dissipated, this implies that, over time, the amount of disorder and randomness in the universe will inevitably grow. This idea of increasing disorder helps us to understand the workings of the second law and why natural processes tend to run only one way: In the direction of increasing entropy, that is, from more ordered states to less ordered states. A glass vase is a very highly ordered arrangement of atoms and molecules that can naturally break into a disordered mass of tiny shards, but it cannot reassemble itself. Similarly, an egg is a highly ordered structure of biological compounds that can fall and become scrambled into a disorderly mess, but the process cannot be reversed. In any system, energy naturally dissipates among all the components of that system, and as this happens entropy grows, the system becomes less structured and more random, and disorder increases.

**RELATED TOPICS**
See also
WHAT IS ENTROPY?
page 140

THE SECOND LAW
page 142

THE CLOSED SYSTEM
page 146

**3-SECOND BIOGRAPHIES**
HERMANN VON HELMHOLTZ
1821–94
German physicist who was first to describe entropy in terms of disorder

JOSIAH WILLARD GIBBS
1839–1903
American mathematician who pioneered the mathematical understanding of entropy, particularly in relation to chemical reactions

**EXPERT**
Leon Clifford

***We expect to see objects become more disordered with time unless something intervenes—any of the processes illustrated here would seem odd reversed.***

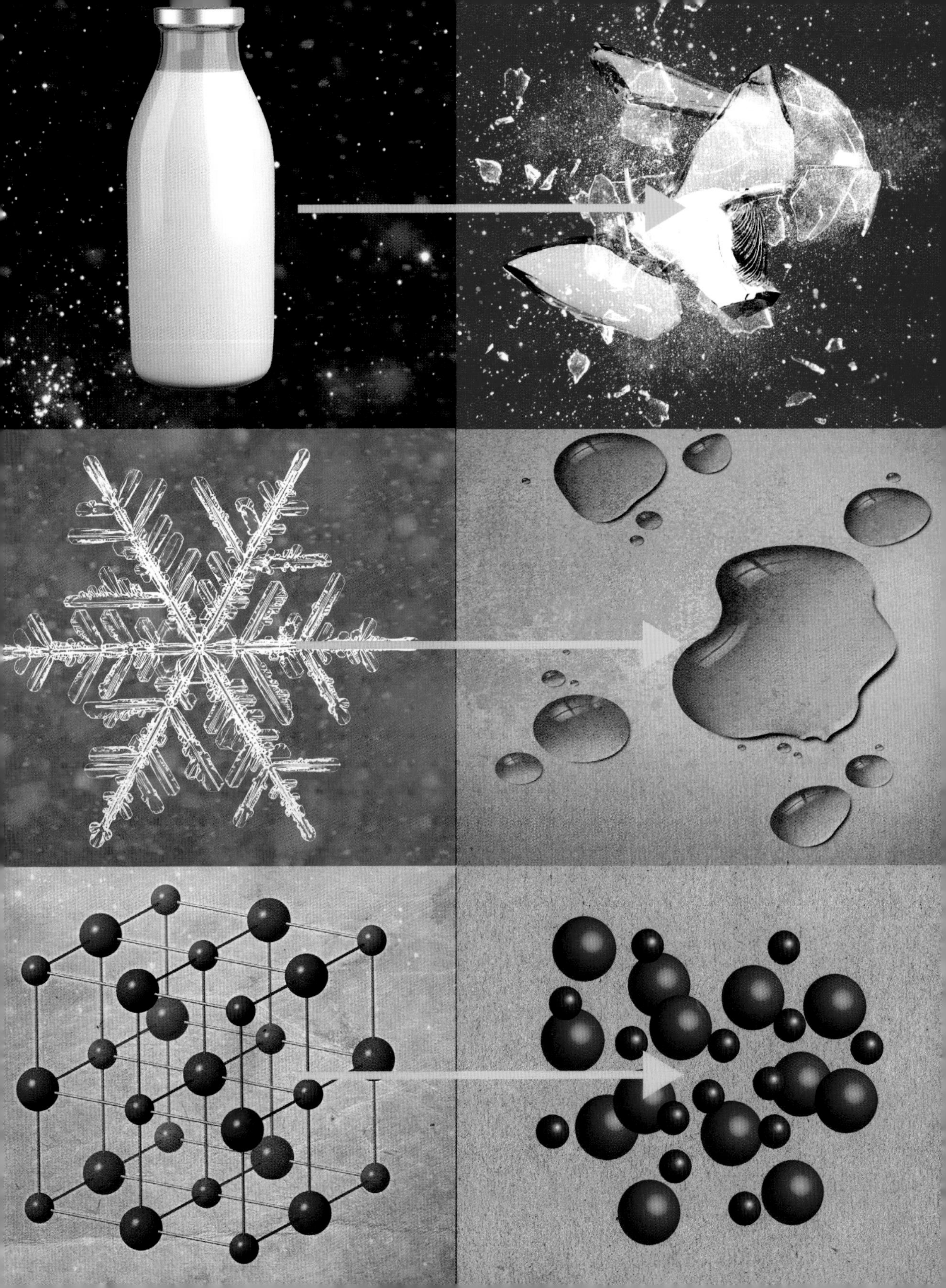

# THE CLOSED SYSTEM

**3-SECOND THRASH**
A closed system can exchange energy but not matter with its surroundings.

**3-MINUTE THOUGHT**
Most refrigerators can be considered closed systems. They work by compressing and expanding a refrigerant fluid in a closed system of tubes. When the refrigerant is expanded and evaporates, it absorbs heat energy from inside the device; on compression it condenses and releases the heat to the outside. The early fridges used poisonous fluids such as ammonia as the refrigerant; if they were not perfectly closed but leaked, they could be (and were) lethal.

Thermodynamics is a kind of energy accountancy: It describes how the budgeting works as energy is moved around and transformed. As with any accounting, this entails keeping a careful track of inflows and outflows. In the simplest situation, no such flows are permitted—neither energy nor matter can be added to or subtracted from the system. This is said to be an isolated system, and generally has a well-defined equilibrium state in which no further change is possible. Our physical universe is thought, virtually by definition, to be like that, although if so it is currently far from attaining its equilibrium state. But some systems can exchange energy, in the form of heat or work, with the surroundings while being unable to exchange matter. These are called closed systems. A chemical reaction inside a sealed vessel is like this; so is a perfectly sealed balloon, which will expand or contract at different temperatures and so perform work on its environment. The former system is the basis of the bomb calorimeter, used to measure the energy change of a chemical process. The latter is like the piston chamber in a steam or combustion engine, where mechanical motion is driven by volume changes in matter due to reaction or heating. Closed systems are important both in fundamental and applied thermodynamics.

**RELATED TOPICS**
See also
CONSERVATION OF ENERGY
page 32

THE SECOND LAW
page 142

INCREASING DISORDER
page 144

**3-SECOND BIOGRAPHIES**
WILLIAM RANKINE
1820–72
Scottish inventor of the Rankine engine, in which heat is delivered to a closed system generally containing water: The basis of steam turbine technology

PIERRE EUGÈNE MARCELLIN BERTHELOT
1827–1907
French chemist who is regarded as the inventor of the bomb calorimeter for measuring heats of combustion

**EXPERT**
Philip Ball

***We can make use of the flow of energy into and out of closed systems to locally overcome the increase of entropy, performing useful tasks.***

# MAXWELL'S DEMON

**3-SECOND THRASH**
Maxwell postulated his demon as a tiny being who, by closely observing molecular motions, could subvert the second law of thermodynamics and engineer a decrease in entropy.

**3-MINUTE THOUGHT**
The problem of Maxwell's demon shows that information itself can be converted to energy. Monitoring and collecting detailed information on the random motions of a particle can create a resource that can be used for "doing work" on the particle. For example, raising it up against gravity. In 2010 a group of Japanese scientists demonstrated this interconversion of energy and information for a plastic bead moving randomly on a "staircase" of electrical force.

There's plenty of energy in the random motions of atoms and molecules, but we can't get at it to do useful work. Or can we? In 1867, James Clerk Maxwell argued that a tiny, sharp-eyed creature (later dubbed a demon) might open and shut a frictionless trapdoor between two compartments to segregate fast-moving (hot) and slow-moving (cool) gas molecules and thus create a temperature difference that could be tapped as an energy source. This, Maxwell knew, contravenes the second law of thermodynamics, which says that temperature differences always get smoothed away as heat flows from hot to cold and entropy increases. It was Maxwell's explicit intention to "pick a hole" in the second law with a situation in which entropy decreases. It took 100 years to figure out why Maxwell's demon wouldn't work. The problem is that the demon would need to store huge amounts of information in its finite brain while observing the molecules. In the 1960s, scientists realized that this information cannot be erased without generating entropy. So even if a demon could induce a decrease in entropy, it would only be fleeting: The need to erase information would more than recoup the lost entropy. The problem of Maxwell's demon links thermodynamics with information theory. It sets a lower limit on the heat that must be dissipated when doing computation.

**RELATED TOPICS**
See also
WHAT IS ENTROPY?
page 140

THE SECOND LAW
page 142

INCREASING DISORDER
page 144

**3-SECOND BIOGRAPHIES**
PETER GUTHRIE TAIT
**1831–1901**
Scottish scientist and expert on thermodynamics, who was the recipient of Maxwell's 1867 letter outlining the demon

ROLF LANDAUER
**1927–99**
German-American physicist who showed in 1961 that erasing one bit of information always dissipates a certain amount of heat, and thus generates entropy

**EXPERT**
Philip Ball

***Maxwell envisaged a being opening and closing a door between compartments, causing entropy to decrease without energy being used.***

1844
Born in Vienna, Austria

1859
Boltzmann's father, Ludwig Georg, dies of tuberculosis, an event that haunted Ludwig for life

1866
Receives his doctorate from the University of Vienna, with a dissertation on the kinetic theory of gases

1869
Appointed professor of mathematical physics at the University of Graz

1872
Publishes the Boltzmann equation, relating entropy to the probabilities of microstates of a system

1888
Accepts and then declines an offer from the University of Berlin—the first sign of a mental crisis

1890
Appointed professor of theoretical physics at the University of Munich

1893
Appointed to the University of Vienna, where he stays—apart from a brief spell at Leipzig

1895
Argues with Ernst Mach and Wilhelm Ostwald about the existence of atoms

1906
Commits suicide in Trieste, Italy

# LUDWIG BOLTZMANN

Ludwig Boltzmann's scientific fame rests on his contributions to statistical mechanics. Born in Vienna, where he studied physics at the university, he took professorships at Graz and Berlin before returning to his hometown to take up the chair in theoretical physics at the university.

Statistical mechanics is the link between the "microscopic" theory of how atoms and molecules move about, and the "macroscopic" theory of their behavior at scales we can see, encoded in the principles of thermodynamics and invoking quantities such as pressure and temperature. In the 1860s and the 1870s Boltzmann, together with James Clerk Maxwell, laid the foundations of statistical mechanical theory by showing how the essentially random motions of countless energetic molecules could lead to steady, predictable relationships between pressure, temperature, and volume.

These ideas are often now assimilated into the broader discipline called statistical physics, which deals with the laws governing huge numbers of interacting "particles" of any kind—they might be atoms or electrons, but could also be sand grains or flocking birds.

As a young man, while studying at the University of Vienna in the 1860s, Boltzmann was introduced to Maxwell's "kinetic theory" of gases. Recognizing that it was neither possible nor particularly meaningful to track all the movements of individual molecules, Maxwell argued that what mattered to their large-scale thermodynamic behavior was the statistical distribution of their properties—their average velocities, say, and how broad the variations about that average value are. Boltzmann was able to show that the distribution of speeds that Maxwell had merely assumed was in fact the inevitable end result of a bunch of molecules moving at random.

Boltzmann's most profound result was a statistical explanation of the second law of thermodynamics, which says that in any spontaneous process in nature the total entropy always increases. He identified entropy $S$—a measure of disorder—with the number $W$ of equivalent arrangements of the constituent particles. The relationship between them, $S=k\log W$, is engraved on his tombstone. Here $k$ is a fundamental constant, now called Boltzmann's constant. Entropy increases, Boltzmann said, simply because that is overwhelmingly more likely: There are many more ways of arranging particles in a more disorderly way than there are in an orderly way, and so random motions are likely to increase the disorder.

Boltzmann's ideas relied on the reality of atoms, for which there was no direct proof in the late nineteenth century. Opposition to the idea contributed to a decline in his mental health in the 1900s, leading him eventually to take his own life. Had he lived, many feel that he could have won the Nobel Prize.

***Philip Ball***

# THE LIFE CYCLE OF THE UNIVERSE

**3-SECOND THRASH**
Our universe started small, dense, and hot and grows larger, colder, and emptier but we do not know how or even whether it will end.

**3-MINUTE THOUGHT**
Our ideas of beginnings and endings are intimately wrapped up with our concept of time and our perception that it flows from the past and into the future. But what if the flow of time we experience is an illusion, as some scientists believe? What if all that has been and all that ever will be actually exists as an infinite block of unchanging spacetime, without any reality to movement through time?

The story of our universe is the story of its energy. Our universe probably began as a subatomic bubble of spacetime that expanded enormously in an instant, due to a process called cosmic inflation that resulted in a seething cauldron of energy. This then exploded outward in what we now think of as the big bang. The universe has been expanding and cooling ever since, according to the laws of thermodynamics. Mass and energy are equivalent and all the particles that make up all the matter in our universe condensed out of the initial energy of the big bang. Although it might seem that the energy of the universe came from nowhere, it is balanced by the gravitational field, acting as if it were negative energy. Our universe may continue expanding and cooling forever; it may collapse in on itself, or it may be part of a cyclic pattern of eternal expansion and contraction. Alternatively, dark energy could rip apart all matter; or cosmic expansion may stretch the fabric of spacetime too far, causing it to break down; or our universe may flip into a lower energy quantum state, leading to what scientists call vacuum decay, which results in a bubble of death expanding outward across the whole cosmos at the speed of light.

**RELATED TOPICS**
See also
GRAVITY
page 38

INFLATION
page 40

DARK ENERGY
page 46

**3-SECOND BIOGRAPHIES**
STEVEN WEINBERG
**born 1933**
American particle physicist who suggested that the different forces of nature were once one and suggested that they condensed separately out of the big bang

JULIAN BARBOUR & MAX TEGMARK
**born 1937 & born 1967**
English physicist and Swedish-born American cosmologist who have both suggested that our experience of the passage of time may be illusory

**EXPERT**
Leon Clifford

***Despite its humble origins, thermodynamics provides our best guide to the eventual possible fate of the universe.***

## RESOURCES

### BOOKS

*A Piece of the Sun*
Daniel Cleary
(The Overlook Press, 2013)

*A Rough Ride to the Future*
James Lovelock
(The Overlook Press, 2015)

*Atmosphere of Hope*
Tim Flannery
(Atlantic Monthly Press, 2015)

*Children of the Sun*
Alfred W. Crosby
(W. W. Norton & Company, 2007)

*Energy: The Subtle Concept*
Jennifer Coopersmith
(Oxford University Press, 2010)

*Energy for Future Presidents*
Richard A. Muller
(W. W. Norton & Company, 2012)

*Ludwig Boltzmann: The Man Who Trusted Atoms*
Carlo Cercignani
(Oxford University Press, 1998)

*Nuclear Power: A Very Short Introduction*
Maxwell Irvine
(Oxford University Press, 2011)

*Renewable Energy*
Godfrey Boyle
(Oxford University Press, 2012)

*Superfuel: Thorium, the Green Energy Source for the Future*
Richard Martin
(St Martin's Press, 2012)

*The Quantum Age*
Brian Clegg
(Icon Books, 2015)

*The Solar Revolution*
Steven McKevitt and Tony Ryan
(Icon Books, 2014)

*Switch*
Cgrus Goodall
(Profile Books Ltd, 2016)

*Thermodynamics for Dummies*
Mike Pauken
(For Dummies, 2011)

## ARTICLES

*Demons, Entropy, and the Quest for Absolute Zero*
Scientific American, March 2011
scientificamerican.com

*In Praise of Lord Kelvin*
Physics World, December 2007
physicsworld.com

*Have We All Been Here Before?*
Focus, August 2008
sciencefocus.com

*Recreating the Sun on Earth*
Focus, April 2014
sciencefocus.com

*The World in 2076: Goodbye Electricity, Hello Superconductivity*
New Scientist, November 2016
newscientist.com

*What is Heat?*
Scientific American, September 1954
scientificamerican.com

## WEBSITES

Environment – Energy
theguardian.com/environment/energy

Electricity
newscientist.com/article-topic/electricity

Energy
scientificamerican.com/energy

Energy and Fuels
newscientist.com/article-topic/energy-and-fuels

Nuclear Power
newscientist.com/article-topic/nuclear-power

Thermodynamics
khanacademy.org/science/physics/thermodynamics

Thermodynamics Advent Calendar
rigb.org/christmas-lectures/supercharged-fuelling-the-future/thermodynamics-2016-advent-calendar

## NOTES ON CONTRIBUTORS

### EDITOR

**Brian Clegg** majored in Natural Sciences, focusing on experimental physics, at the University of Cambridge. After developing high-tech solutions for British Airways and working with creativity guru Edward de Bono, he formed a creativity consultancy advising clients ranging from the BBC to the Met Office. He has written for *Nature*, *The Times*, and the *Wall Street Journal*, and has lectured at Oxford and Cambridge universities and the Royal Institution. He is editor of the book review site popularscience.co.uk, and his own published titles include *A Brief History of Infinity*, *How to Build a Time Machine*, *The Reality Frame*, and *Are Numbers Real?*

### FOREWORD

**Professor Jim Al-Khalili OBE** is a physicist, author, and broadcaster based at the University of Surrey. He received his PhD in theoretical nuclear physics in 1989 and has published over a hundred research papers on quantum physics. His many popular science books have been translated into 26 languages. He is a recipient of the Royal Society Michael Faraday medal and the Institute of Physics Kelvin Medal. Jim is a regular contributor to radio and television science programs. In 2016 he received the inaugural Stephen Hawking medal for science communication.

### CONTRIBUTORS

**Philip Ball** is a freelance writer, and was an editor for *Nature* for more than 20 years. Trained as a chemist at the University of Oxford, and as a physicist at the University of Bristol, he writes regularly in the scientific and popular media, and has authored books including *$H_2O$: A Biography of Water*, *Bright Earth: Art and the Invention of Color*, *The Music Instinct: How Music Works and Why We Can't Do Without It*, and *Curiosity: How Science Became Interested in Everything*. His book *Critical Mass: How One Thing Leads to Another* won the 2005 Aventis Prize for Science Books and his latest book is *The Water Kingdom: A Secret History of China*. He has been awarded the American Chemical Society's Grady-Stack Award for interpreting chemistry to the public, and was the inaugural recipient of the Lagrange Prize for communicating complex science.

**Leon Clifford** is managing director of science communications consultancy Green Ink Publishing Services ltd. Leon has a BSc in physics-with-astrophysics and is a member of the Association of British Science Writers. He worked for many years as a journalist covering science, technology, and business issues with articles appearing in numerous publications including *Electronics Weekly*, *Wireless World*, *Computer Weekly*, *New Scientist*, and *The Telegraph*. Leon is interested in all aspects of physics—particularly climate science, astrophysics, and particle physics.

**Simon Flynn** worked in publishing for fifteen years and is now a science teacher in London. He is author of *The Science Magpie: A Hoard of Fascinating Facts, Stories, Poems, Diagrams, and Jokes Plucked from Science and Its History,* which Physics World chose as one of its top 10 books of 2012, and a contributor to *What if Einstein Was Wrong?* and *30-Second Newton.*

**Sharon Ann Holgate** is a freelance science writer and broadcaster with a doctorate in physics. She has written for newspapers and magazines including *Science* and *New Scientist*, and presented on BBC Radio 4 and the BBC World Service. She was coauthor of *The Way Science Works*, a children's popular science book shortlisted for the 2003 Junior Prize in the Aventis Prizes for Science Books, and wrote the textbooks *Understanding Solid State Physics* and *Outside the Research Lab – Volume 1: Physics in the Arts, Architecture and Design*. In 2006, Sharon Ann won Young Professional Physicist of the Year for her work communicating physics.

**Andrew May** is a technical consultant and freelance writer on subjects ranging from astronomy and quantum physics to defense analysis and military technology. After reading Natural Sciences at the University of Cambridge in the 1970s, he went on to gain a PhD in Astrophysics from the University of Manchester. Since then he has accumulated more than 30 years' worth of diverse experience in academia, the scientific civil service, and private industry.

# INDEX